T0211923

Sensory Organ Replacement and Repair

Sensory Organ Replacement and Repair
Gerald E. Miller

ISBN: 978-3-031-00484-1   paperback   Miller
ISBN: 978-3-031-01612-7   ebook   Miller

DOI: 10.1007/978-3-031-01612-7

A Publication in the Springer series
SYNTHESIS LECTURES ON BIOMEDICAL ENGINEERING
Lecture #3

First Edition

Sensory Organ Replacement and Repair

Gerald E. Miller
Virginia Commonwealth University

SYNTHESIS LECTURES ON BIOMEDICAL ENGINEERING #3

ABSTRACT

The senses of human hearing and sight are often taken for granted by many individuals until they are lost or adversely affected. Millions of individuals suffer from partial or total hearing loss and millions of others have impaired vision. The technologies associated with augmenting these two human senses range from simple hearing aids to complex cochlear implants, and from (now commonplace) intraocular lenses to complex artificial corneas. The areas of human hearing and human sight will be described in detail with the associated array of technologies also described.

KEYWORDS

Human hearing, Audiology, Hearing Aids, Cochlear Implants, Human vision, Intraocular lens, Cataract surgery, Artificial Cornea, Corneal Transplant

Contents

Sensory Organ Replacement and Repair

The senses of human hearing and sight are often taken for granted by many individuals until they are lost or adversely affected. Millions of individuals suffer from partial or total hearing loss and millions of others have impaired vision. The technologies associated with augmenting these two human senses range from simple hearing aids to complex cochlear implants, and from (now commonplace) intraocular lenses (IOLs) to complex artificial corneas. The areas of human hearing and human sight will be described in detail with the associated array of technologies also described.

1   HEARING AIDS

1.1   Anatomy of the Ear and Human Hearing

The human ear is a complex arrangement of mechanical and neurological components that are designed to interpret a multifaceted sound pressure waveform into its individual frequency components and then reconstitute these components together into what we recognize as blended sound. The human ear is shown in Figure 1.

The ear is composed of three sections: the outer ear (also known as the ear canal), the middle ear, and the inner ear. The outer ear is shown in greater detail in Figure 2.

The outer ear is designed to channel sound from a complete hemisphere into a megaphone-shaped channel that is wide at the outside end and funnels down toward the inside end. As was noted above, most sounds are complex waveforms consisting of pressure waves of varying amplitudes and frequencies mixed together into sounds and words that we recognize in our daily lives. A typical sound waveform of spoken words is shown in Figure 3 with the frequency spectrum of those words shown in Figure 4.

The array of pressure waveforms (Figures 3 and 4) results from a series of anatomical elements including the trachea, lungs, vocal cords, tongue, and mouth, as is seen in Figure 5.

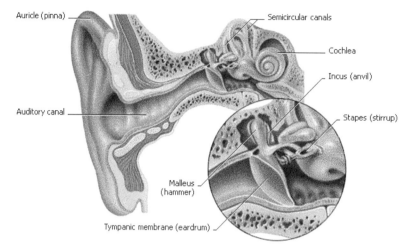

FIGURE 1: Anatomy of the human ear.

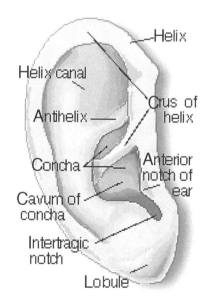

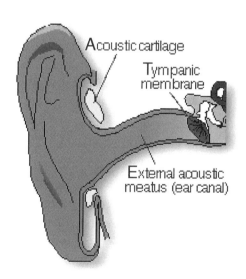

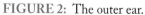

FIGURE 2: The outer ear.

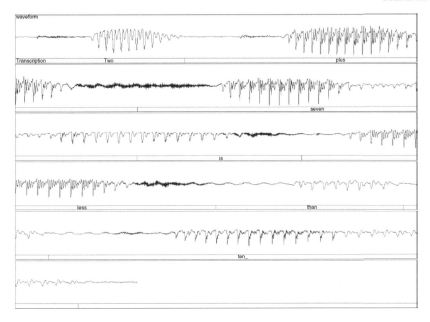

Two plus seven is less than ten

FIGURE 3: Speech waveform of spoken words indicating a breadth of amplitudes and frequencies.

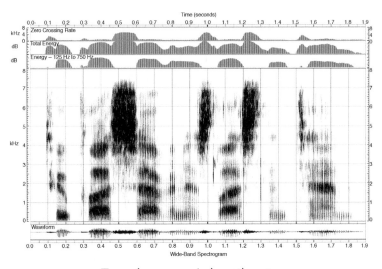

Two plus seven is less than ten

FIGURE 4: Frequency spectrum of spoken words indicating a range of embedded frequencies.

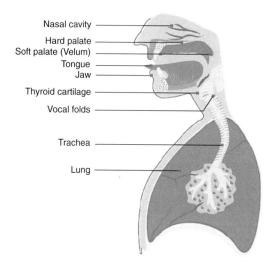

FIGURE 5: Anatomical structures associated with speech production.

The complex waveforms associated with speech and many sounds are channeled into the outer ear and, as noted above, funneled toward the inside, which culminates with the tympanic membrane, also known as the eardrum. This membrane vibrates with the same complex arrangement as the complex sound waveforms that enter the ear canal.

From the tympanic membrane, the vibrations associated with the sound pressure waveform are then transmitted into the middle ear, which consists of three tiny bones called the ossicles (the malleus, incus, and stapes). These three bones are sometimes given the colloquial names hammer, anvil, and stirrups, which describe their shapes. They amplify the sound and send it through the entrance to the inner ear (oval window) and into the fluid-filled hearing organ known as the cochlea. The middle ear appears to be an overly complex structure of three bones that connect one membrane (the tympanic membrane or eardrum) to the oval window leading to the inner ear. However, there are two important elements to the middle ear. Firstly, the bones not only transmit the complex sounds from the tympanic membrane to the oval window, but also amplify the pressure due to the area difference between the two membranes and the size of the bones that attach to these membranes. In this fashion, the middle ear is similar to an impedance-matching device. Secondly, the middle ear is responsible for balance. If there is a significant pressure difference between the two ears, then one's balance is adversely affected.

The sound from the middle ear is transmitted into the inner ear known as the cochlea. It is here that the complex sound waveforms of varying frequencies and amplitudes are separated into individual components. This is accomplished by means of the resonant frequency for a material based upon its geometry. Thicker materials vibrate naturally at lower frequencies, while thinner

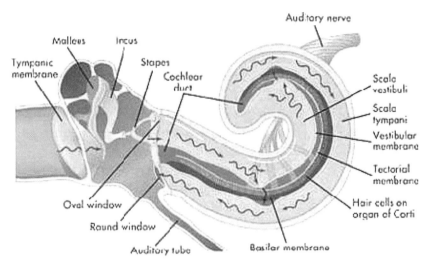

FIGURE 6: Human cochlea with hair cells and auditory nerve.

materials vibrate at higher frequencies. This is why bass speakers are larger than treble speakers in a stereo system. The cochlea is configured as a coiled triangular tissue with a wide thicker end and a narrow thinner end with a continual taper along its length. As complex sound enters into the cochlea, each relevant section of the cochlea would vibrate at its resonant frequency based upon its thickness. Thus, the complex sound is then separated into its individual frequency elements. The cochlea is thus a reverse auditory sound mixer. Along the cochlea, there are thousands of hair cells that are connected to tiny nerves. As the various sections of the cochlea vibrate at each resonant frequency at an amplitude dependent on the incoming sound amplitude (at that frequency), so do the attached hair cells. These hair cells excite their associated nerves. These nerves are then connected *en masse* to the auditory nerve or eighth cranial nerve, which sends the recombined electrical signal to the brain. The brain translates these impulses into what we experience as sound. It is unusual that the cochlea separates sounds into their individual components, converts them from vibrations into electrical signals, and then reconstitutes them into a blended sound, similar to its original construct. Figure 6 indicates the cochlea with hair cells leading to the auditory nerve.

The wide variety of sounds that we routinely hear vary from a whisper to a roar with pressure variations covering several orders of magnitude. To provide some uniform measurement system, a logarithmic sound scale has been established in units known as decibels. A decibel scale for many common sounds is shown in Figure 7.

The frequency range of human hearing is 20–20 000 Hz with the lowest end being more of a vibratory feeling than true auditory hearing. The range of human hearing as compared to other species is shown in Figure 8.

Decibel	Sound
0–9	One hand moving in air
10–19	Softest sound
20–29	Softest whisper
40–49	Quiet office (no typing)
50–59	Average home with A/C
60–69	Normal conversations
70–79	Living room music, radio, busy traffic
80–89	Food blender, noisy restaurant
90–99	Motorcycle at 25 ft, subway
100–109	Airplanes 1000 ft away
110–119	Rock concerts
>120	Jets, rockets

FIGURE 7: Decibel scale representing common sounds at various amplitudes.

Hearing is a very powerful sense—far more powerful than most people realize. Hearing allows individuals to determine where a sound is coming from, even if it is behind them or they cannot see the source of the sound. In addition, in a noisy environment, individuals can selectively filter out "extraneous" sound and concentrate on one voice or sound source. Try this

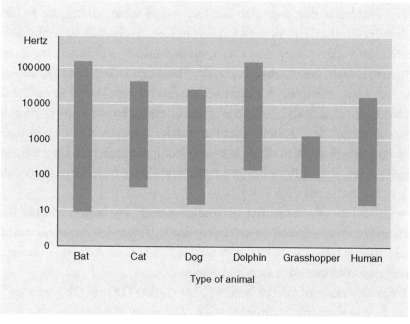

FIGURE 8: Frequency range of human hearing compared to other species.

experiment: In a crowded room, close your eyes. Then try to concentrate on one person's voice and determine where that person is merely from the sounds reaching your ear. As such, hearing is truly three dimensional and, of more importance, can be selective among several blended sounds. This is important when one has a hearing impairment that requires augmentation with a hearing aid or cochlear implant. One may regain sound amplitude and frequency processing, but loses the spatial resolution feature that is such a powerful aspect of normal hearing.

1.2   Hearing Loss

When the problem is in the inner ear, a *sensorineural* hearing loss occurs. Sensorineural hearing loss is the most common type of hearing loss. More than 90% of all hearing aid wearers have sensorineural hearing loss. The most common causes of sensorineural hearing loss are age-related changes, noise exposure, inner ear blood circulation, inner ear fluid disturbances, and problems with the hearing nerve. *Conductive* hearing loss occurs when sound is not conducted efficiently through the ear canal, eardrum or the tiny bones of the middle ear, resulting in a reduction of loudness of sound. Conductive loss may result from earwax blocking the ear canal, fluid in the middle ear, middle ear infection, obstructions in the ear canal, perforations (hole) in the eardrum, or disease of any of the three middle ear bones. People with conductive hearing loss may notice that their ears seem to be full or plugged. They may speak softly because they hear their own voice loudly. Crunchy foods, such as celery or carrots, seem very loud to the person with a conductive hearing loss and this person may have to stop chewing to hear what is being said. All conductive hearing losses should be evaluated by an audiologist and a physician to explore medical and surgical options.

Audiometry is the testing of a person's ability to hear various sound frequencies. The test is performed with the use of electronic equipment called an *audiometer*. This testing is usually administered by a trained technician called an *audiologist*. Audiometry testing is used to identify and diagnose hearing loss. The equipment is used in health screening programs, for example, in grade schools, to detect hearing problems in children. It is also used in the doctor's office or in hospital's audiology department to diagnose hearing problems in children, adults, and the elderly. With correct diagnosis of a person's specific pattern of hearing impairment, the right type of therapy, which might include hearing aids, corrective surgery, or speech therapy, can be prescribed.

The person being tested wears a set of headphones that blocks out other distracting sounds and delivers a test tone to one ear at a time with the amplitude slowly increasing. When the patient hears the sound of a tone, he holds up a hand or finger to indicate that the sound is detected. The audiologist lowers the volume and repeats the sound until the patient can no longer detect it. This process is repeated over a wide range of tones or frequencies from very

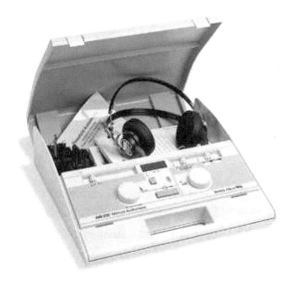

FIGURE 9: Small desktop audiometer.

deep, low sounds, like the lowest note played on a tuba, to very high sounds, like the pinging of a triangle. Each ear is tested separately. It is not unusual for levels of sensitivity to sound to differ from one ear to the other. The results of the audiometry test may be recorded on a grid or graph called an audiogram. This graph is generally set up with low frequencies or tones at one end and high ones at the other end, much like a piano keyboard. Low notes are graphed on the left and high notes on the right. The graph also charts the volume of the tones used, from soft, quiet sounds at the top of the chart to loud sounds at the bottom. Hearing is measured in a unit called decibels. Most of the sounds associated with normal speech patterns are generally spoken in the range of 20–50 dB. An adult with normal hearing can detect tones between 0 and 20 dB.

A typical desktop audiometer is shown in Figure 9 with a computer-driven system shown in Figure 10.

Some hospitals and schools utilize an acoustic chamber type audiometer that further isolates interfering sound. This chamber is sometimes called an anechoic chamber and is shown in Figure 11.

The audiograms noted above can be computer generated as seen in Figure 12 or hand drawn as with a desktop system as seen in Figure 13.

As can be seen from the audiograms given in Figures 12 and 13, there are data points at various frequencies for both ears. Normal hearing is designated by a fairly parallel set of points for each ear as is shown in Figure 14.

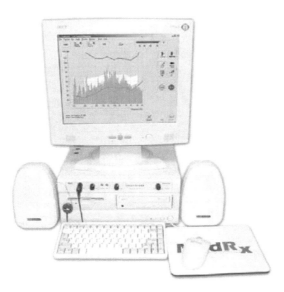

FIGURE 10: Computer-controlled audiometer.

FIGURE 11: Acoustic chamber type audiometer.

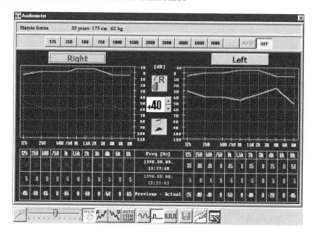

FIGURE 12: Computer-generated audiogram.

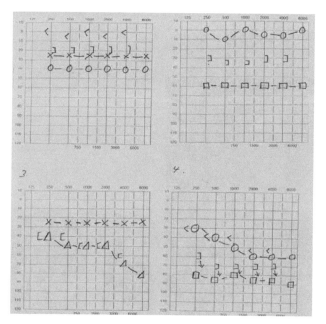

FIGURE 13: Hand-drawn audiogram from a desktop audiometer.

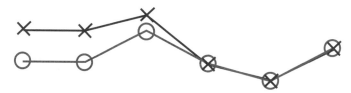

FIGURE 14: Normal bilateral hearing with fairly parallel data points at all frequencies.

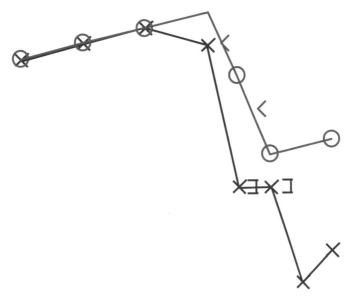

FIGURE 15: Upper frequency hearing loss in both ears.

Elderly individuals or those who have been repeatedly exposed to loud sounds often lose hearing at the upper frequencies as shown in Figure 15. Still other patients may have hearing loss in mid-frequency ranges due to childhood illnesses or recurring tympanic membrane problems, as is seen in Figure 16.

There are sometimes hearing losses in only one ear, typically due to recurring infection in one ear or damage to the outer or middle ear in one ear. When both ears are affected, the patient may utilize hearing aids in both ears, although that is not required. Obviously, when only one ear is affected, a single hearing aid may be needed. Figure 17 depicts one-sided hearing loss with a patient who has a recurring right ear infection.

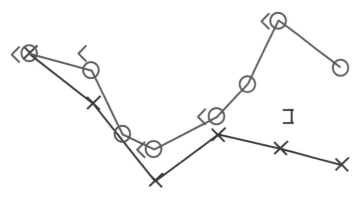

FIGURE 16: Mid-range hearing loss in both ears.

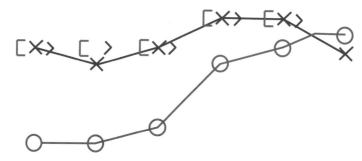

FIGURE 17: One-sided hearing loss at most frequencies due to a recurring right ear infection.

1.3 Hearing Aid Technologies

Hearing aids are actually miniature sound systems with a microphone, speaker, audio amplifier, PA and associated electronics. As with many sound systems, the amplitude of the amplified sound can be adjusted as can the frequency range. In addition, as with many stereo systems, issues such as total harmonic distortion, signal to noise ratio, overall gain, and other signal processing factors are also elements in hearing aids. Since any individual patient has an individualized hearing loss, which can be frequency and amplitude dependent, there are settings for gain at varying frequencies that are preset for each patient, and then are fixed. The patient can have control over the overall gain and whether the unit is turned on or off, but cannot normally set the gain at individual frequencies, as these are internally set within the casing of the hearing aid within the electronics. A small dime-sized battery, which is normally on a pivot holder to allow for ease of replacement, powers the device. A typical hearing aid is shown in Figure 18 that depicts an ear mold (not used in all types of hearing aids), volume control, and on–off switch.

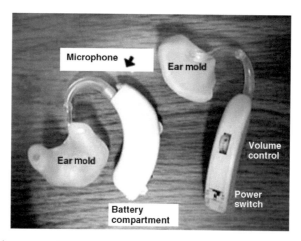

FIGURE 18: Typical hearing aid configuration with volume control and on–off switch.

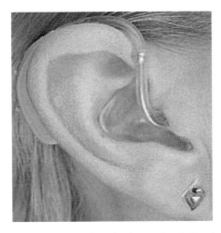

FIGURE 19: BTE hearing aid with clear tube fitting into the ear canal.

There are four styles or configurations of hearing aids. These include the "behind-the-ear" (BTE) style, the "in-the-canal" (ITC) style, and the "completely-in-the-canal" (CIC) style. The BTE hearing instruments are extremely flexible for all types of hearing loss.

The hearing device is housed within a curved shell that sits behind each ear and delivers sound through a clear tube, as is seen in Figure 19. The clear tube fits into a mold that has been customized to comfortably fit inside each ear.

The in-the-ear (ITE) hearing instruments are very easy to operate, even if the user has poor dexterity. The hearing device is housed within a custom-made shell that fits comfortably inside each ear and delivers sound directly to the ear. This style is shown in Figure 20 and is

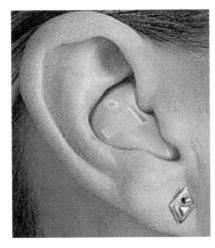

FIGURE 20: ITE style of hearing aid with the device easily accessible and removable.

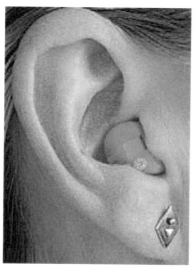

FIGURE 21: ITC style hearing aid with the device positioned further into the canal.

one of the most common styles of hearing aids, particularly for the elderly or for individuals wearing glasses who cannot easily utilize the BTE style.

The ITC hearing instruments can barely be seen and are very easy to operate, even if the user has poor dexterity. The hearing device is housed within a custom-made shell that fits comfortably inside each ear canal and delivers sound directly to the ear. However, this style is not as accessible as the ITE style. This version is seen in Figure 21 and is often utilized by younger patients as opposed to elderly patients.

The CIC hearing instruments are virtually invisible to others. The hearing device is housed in a tiny shell that fits comfortably and completely into each ear canal. The device is removed from the ear canal by pulling a tiny cord. Where these miniature instruments are both powerful and cosmetically appealing, some features such as manual volume control are not available simply because the devices are so small. These types of devices are worn almost exclusively by younger patients who wish to avoid the appearance of a hearing aid. This style is shown in Figure 22.

Hearing aid batteries are small dime-sized devices housed within a pivoting holder within the shell of the device. As is the case with many electronic devices, there are many types of hearing aids. However, in order to avoid confusion for a health care device, hearing aid batteries are color coded, so that it is easy to purchase the correct replacement battery, as is seen in Figure 23.

Typically, batteries last 7–14 days based on 16 h per day use cycle. Batteries are inexpensive, costing less than a dollar each. Generally, the smaller the battery size, the shorter the battery

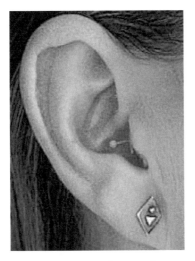

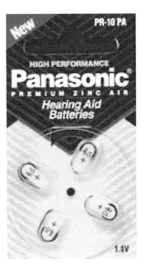

FIGURE 22: CIC style with a protruding cord to remove the device.

life. The sizes of hearing aid batteries are listed below along with their standard numbers and color codes:

Size 5: Red
Size 10 (or 230): Yellow
Size 13: Orange
Size 312: Brown
Size 675: Blue

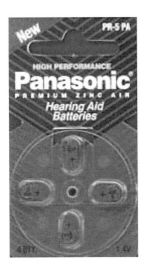
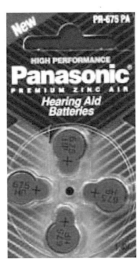
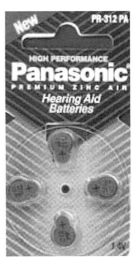

FIGURE 23: Color-coded hearing aid batteries with easy punch out sections and easily seen plus/minus poles.

Today's hearing aid batteries are "zinc-air." Because the batteries are air activated, a factory-sealed sticker keeps them "inactive" until you remove the sticker or push the battery from within its sealed package. Once the sticker is removed from the back of the battery, oxygen in the air contacts the zinc within the battery, and the battery is activated. Zinc-air batteries have a "shelf life" of up to 3 years when stored in a cool, dry environment. Storing zinc-air hearing aids in the refrigerator has no beneficial effect on their shelf life. In fact, quite the opposite may happen. The cold air may actually form little water particles under the sticker.

Hearing aid electronics may be analog or digital. The analog versions may be programmable, but the digital versions are all programmable. Most people recognize that digital hearing aids provide better quality to the user; however, some people still use analog hearing aids for the simple fact that digital models are considerably more expensive. Analog hearing aids usually cost anywhere from $200 to $1000 and digital hearing aids usually cost between $800 and $2000. This means that digital hearing aid models can potentially cost more than twice as much as their analog counterpart. In general, digital hearing aids provide a higher quality, more "realistic" sound with the ability to filter out ambient, nonspeech sounds. In addition, they are more easily programmable in terms of frequency ranges to be augmented and/or depressed and are also smaller in size than their analog counterparts.

There are various auxiliary devices used by individuals with hearing impairments who also use hearing aids. These include modified telephones, alarm clocks, television audio controls and interfaces, and radio/stereo interfaces. One vital component to hearing aids is the use of a telecoil (T-coil), which provides an interface to a telephone while avoiding feedback to the hearing aid. In some cases, a telephone earpad is placed over the earpiece such that the incidence of feedback to the hearing aid is also reduced, as is shown in Figure 24.

In most cases, an amplified telephone is used by individuals with hearing aids to allow for selectively increased amplification of the spoken voice over the phone. The telephone bandwidth is 300–3000 Hz. Amplified telephones are T-coil compatible, often have tone controls, may have audio jacks for headsets or attachment to cochlear implants (to be discussed below), and they amplify the speech up to 40 dB, which is considerable. The ringer is also adjustable up to 95 dB, which is the sound level equivalent to a motorcycle or subway. In addition, there is typically a large flashing visual display for the ringer to further alert a hearing impaired individual to an incoming call. There are third-party add-ons to standard telephones that allow for adjustable volume control and a visual display as shown in Figure 25.

In order to avoid outside interfering sounds and also to provide adjustable volume control, those with impaired hearing or who wear hearing aids sometimes employ an infrared wireless system that can attach to a standard television as shown in Figure 26.

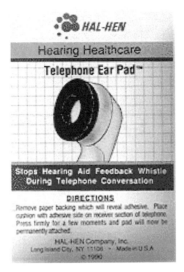

FIGURE 24: Telephone earpad to allow the hearing aid to be more distant to the telephone speaker and reduce interference and feedback.

As was noted above, special alarm clocks can be purchased that allow for adjustable volume increase up to 40 dB and have a large (sometimes projecting) visual display when the alarm sounds.

The Federal Trade Commission (FTC) is responsible for monitoring the business practices of hearing aid dispensers and vendors. The FTC can take action against companies that mislead or deceive consumers. Such companies may use misleading sales and advertising

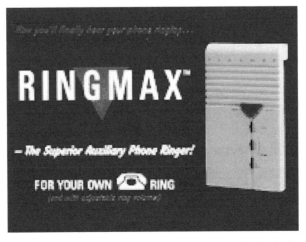

FIGURE 25: Adjustable volume control as an attachment to a standard telephone.

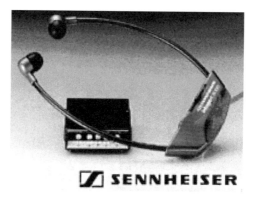

FIGURE 26: Infrared wireless headset for television listening and interface.

practices, giving inaccurate information about hearing loss, hearing aid performance, refund policies, or warranty coverage. The law further requires companies offering warranties to fully disclose all terms and conditions of their warranties.

The Food and Drug Administration (FDA) enforces regulations that deal specifically with the manufacture and sale of hearing aids. According to the FDA, the following conditions must be met by all dispensers before selling a hearing aid:

1. Dispensers must obtain a written statement from the patient, signed by a licensed physician. The statement must be dated within the previous 6 months, state that the patient's ears have been medically evaluated, and that the patient is cleared for fitting with a hearing aid.

2. A patient of age 18 years or older can sign a waiver for a medical examination, but dispensers must avoid encouraging the patient to waive the medical evaluation requirement. Dispensers also must advise the patient that waiving the examination is not in his best health interest.

3. Dispensers must advise patients who appear to have a hearing problem to consult promptly with a physician.

4. The FDA regulations also require that an instruction brochure be provided with the hearing aid that illustrates and describes its operation, use, and care. The brochure must list sources for repair and maintenance, and include a statement that the use of a hearing aid may be only a part of a rehabilitative program.

The FDA Web site that notes standards for hearing aids is at http://www.accessdata. fda.gov/scripts/cdrh/cfdocs/cfStandards/Detail.CFM?STANDARD_IDENTIFICATION_

NO=14730. Hearing aids are regulated by the FDA within the Center for Devices and Radiological Health, which can be accessed at http://www.fda.gov/cdrh/.

Numerous studies have been performed regarding hearing aid use, design, and evaluation. Studies involving the use of hearing aids by the elderly have been published by Cohen-Mansfield and Infeld (2005), and van Hooren *et al.* (2005), among others. Studies regarding the evaluation of hearing aid electronic design include those by Bentler (2005), Lewis *et al.* (2005), Moore *et al.* (2005), and Ricketts and Hornsby (2005), among others. Still other studies have examined the methods of fitting hearing aids in patients or evaluation of first-time users, which include Aarts and Caffee (2005), Gustav Mueller (2005), Gustav Mueller and Bentler (2005), Killion and Gudmundsen (2005), Reber and Kompis (2005), Reese and Hnath-Chisolm (2005), and Uriarte *et al.* (2005), among others. General reviews regarding the use of hearing aids include those by Vuorialho *et al.* (2005).

2   MIDDLE EAR REPLACEMENT

2.1   Introduction

Otosclerosis is a condition that affects hearing as a result of hardening of a bone or bones in the middle ear. The hearing loss associated with this disease is called a conductive hearing loss because the extra bone growth around the middle ear bones prevents sound from being conducted into the inner ear in a normal way. Otosclerosis is inherited and it tends to run in families although you may not know who in your family passed it on to you. About 80% of people with otosclerosis will have the disease in both ears. If you have otosclerosis, it is estimated that there is less than a one in four chance of passing it on to your children.

Sound vibrations that reach the eardrum are usually relayed to the inner ear by way of three small bones in the middle ear. These tiny bones, called the hammer (malleus), anvil (incus), and stirrup (stapes), act as a kind of transformer to change sound waves into liquid waves in the inner ear. The stapes bone is the final link in the hearing chain of bones and is the bone most often affected by otosclerosis. When the stapes hardens, a conductive hearing loss occurs. As the hardening continues over time, your hearing worsens. When the hardening spreads to the inner ear, a sensorineural hearing loss occurs. A sensorineural hearing loss is a nerve hearing loss that is usually permanent and can only be helped with a hearing aid.

Hearing loss is the most common symptom of otosclerosis. Some people also have tinnitus, a ringing noise in the head or ear, and almost half of all people with this disease have dizziness. A woman with otosclerosis that becomes pregnant might find that her hearing loss becomes worse.

Unfortunately, there is no medicine that will help or stabilize the hearing loss in people who have otosclerosis. For many people, however, surgery can help or even overcome the hearing

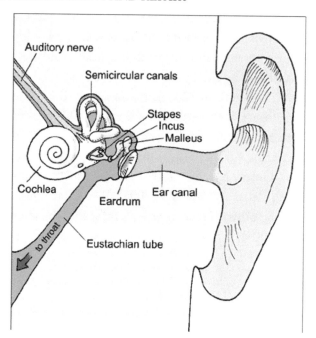

FIGURE 27: Bones of the middle ear.

loss they have. There are several surgery techniques used to correct the hearing loss associated with otosclerosis. A stapedectomy or stapedotomy is usually recommended. These operations are usually done in a hospital or surgery center with a local anesthetic. After your ear has been numbed, it is cleaned and then a cut is made down inside the ear canal and the eardrum is lifted up to uncover the middle ear. The diseased stapes bone is removed, the inner ear is sealed with tissue, and a new stapes is inserted. Sometimes a laser is used to open the bottom of the stapes and a pistonlike stapes replacement is used. It is also possible that a replacement malleus and/or incus is required. Figure 27 depicts the middle ear among the various anatomical features of the human ear.

2.2   Technology and Replacement Components

When the incus is eroded, broken, or absent, the ossicular chain is reconstructed with an incus replacement prosthesis. The one depicted in Figure 28 is a cylinder with a notch that fits under the handle of the malleus and a circular groove that sits on the head of the stapes.

When both the incus and the malleus are eroded or absent, the ossicular chain is re-constructed with a partial ossicular replacement prosthesis (PORP). The one shown in the Figure 29 is a cylinder with a circular flange that fits under the drum. A piece of cartilage is

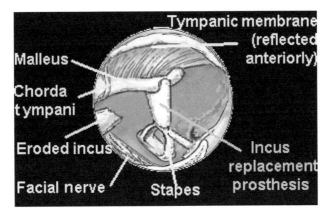

FIGURE 28: Incus replacement with artificial incus.

removed from the tragus and inserted between the prosthesis and the inner surface of the drum to minimize rejection.

When the incus and arch of the stapes are eroded, or when the malleus, incus, and arch of the stapes are absent, the ossicular chain is reconstructed with a total ossicular replacement prosthesis (TORP). The one depicted in Figure 30 has a circular flange that fits under the drum and a slender shaft that is placed over the footplate of the stapes. Here again, a piece of cartilage is inserted between the drum and the flange.

Surgical replacement of the malleus is shown in Figure 31 and surgical replacement of the incus in Figure 32.

The actual size of an artificial stapes, malleus, or incus is shown in Figure 33 as compared to a common dime.

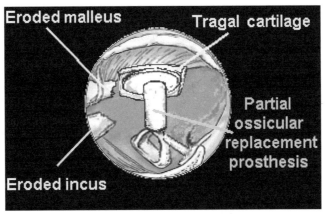

FIGURE 29: PORP replacement of both the incus and the malleus.

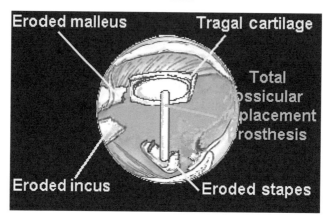

FIGURE 30: TORP replacement of both the stapes and the incus.

2.3   History of Ossicle Surgery and Replacement

Otosclerosis surgery has developed through three eras. The mobilization era began in the late 1800s when Kessel attempted stapes mobilization without ossicular chain reconstruction in cases where it was noted to be fixed (Heermann, 1969). Later, Jack removed the stapes, leaving the oval window open (Schuknecht, 1968). Both techniques allowed increased transmission of sound through the oval window but did not use middle ear amplification structures. Furthermore, fatal cases of meningitis from intraoperative exposure of perilymph to bacteria occurred, and

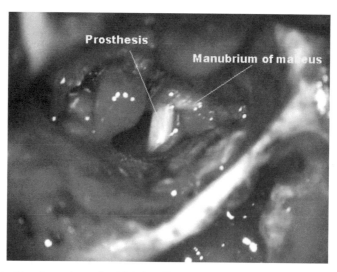

FIGURE 31: Surgical implantation of artificial malleus. Image is significantly magnified with the actual prosthesis of the size of a letter "i" in an 800 × 600 display.

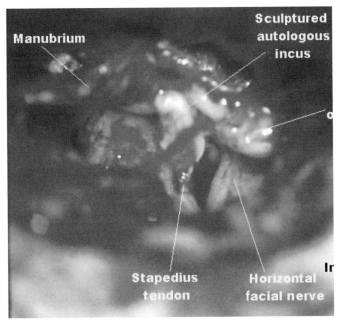

FIGURE 32: Surgical implantation of an artificial incus. Image is significantly magnified.

any gains in hearing frequently were temporary because any remaining stapes footplate often refixed.

The fenestration era began in 1923, when Holmgren created a fistula in the horizontal semicircular canal and sealed it immediately with periosteum (Shea, 1998). This procedure allowed sound conduction preferentially through the fistula, rather than the ossicular chain. Sourdille popularized the procedure when his three-stage technique was widely published during the 1930s (Shea, 1998). Lempert developed a one-stage technique for horizontal semicircular

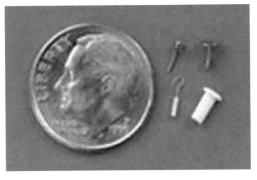

FIGURE 33: Actual size of artificial ossicle (middle bone) components.

fenestration, which went on to gain worldwide acceptance after it proved to enhance hearing (Lempert *et al.*, 1956). Results, however, were short-lived because the fenestra often resealed with bone.

The stapedectomy era began before the fenestration era closed. Rosen revisited stapes mobilization in 1952 (Shea, 1998). Later, Shea removed the stapes, sealed the oval window with an autograft vein wall, and then reconstructed the sound-conducting mechanism with an artificial prosthesis (House *et al.*, 1960; Shea, 1976).

This technique gained wide acceptance and has been improved since inception. In the 1970s, Myers conducted stapedotomy using a piston prosthesis (Myers *et al.*, 1970; Myers and Myers, 1968). In the early 1980s, Perkins began using the laser for stapedotomy in a procedure in which a small hole is made in the footplate, as opposed to complete or subtotal removal (Perkins, 1980). Several techniques and approaches are commonly used today, with largely excellent results. A few challenges remain, such as those patients enduring sensorineural hearing loss and unsteadiness, but many think that surgical treatment for otosclerosis has reached perfection.

Evidence has recently mounted that the measles virus plays an important role in gene activation of otosclerosis (Ferlito *et al.*, 2003; Karosi *et al.*, 2004; Karosi *et al.*, 2005; Niedermeyer *et al.*, 2001). This hypothesis is supported by a declining incidence of otosclerosis since measles vaccinations became widespread.

3   COCHLEAR IMPLANT

3.1   Introduction

Individuals who have extreme sensorineural hearing loss, are born completely deaf, acquire deafness through illness or injury, or who cannot be adequately treated with the use of a hearing aid or artificial ossicle are candidates for a cochlear implant. The cochlear implant is a prosthetic replacement for the inner ear (cochlea) and is appropriate only for people who receive minimal or no benefit from a conventional hearing aid. The cochlear implant bypasses damaged parts of the inner ear and electronically stimulates the hair cells and adjacent nerves within the cochlea. Part of the device is surgically implanted in the skull behind the ear and tiny wires are inserted into the cochlea at set intervals depending on the number of channels or number of frequency bands to excite. The other part of the device is external and has a microphone, a speech processor (that converts sound into electrical impulses), and connecting cables. It is battery powered, adjustable, and expensive.

A surgical procedure places the implant on the outside of the skull and under the skin just above and behind the ear. An electrode array attached to the implant is inserted into the cochlea. Within a few weeks of the surgery, the user is fitted with external devices—a microphone, processor, and transmitter coil. Sound waves are received by the microphone that

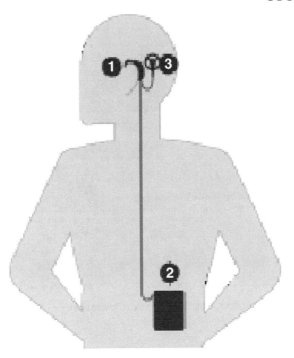

FIGURE 34: Schematic view of a cochlear implant with (1) BTE speech processor or (2) on-the-body speech processor. Either is connected to (3) the connector to the electrode array.

is located either above the ear (Figure 34, point 1) or on the transmitter coil (Figure 34, point 3). The signal from the microphone is sent to the speech processor, which comes in two designs. It may be either a BTE model, which looks like a hearing aid (Figure 34, point 1), or a body-worn device (BWD) that rests on the belt (Figure 34, point 2). In either case, the speech processor deciphers the sound and decides how it will be presented to the ear by means of which electrodes (representing frequencies/regions of the cochlea) will be excited. There are many ways that a processor can translate sound into an electronic code, each of which is called a speech strategy. New speech strategies are being constantly developed to help cochlear implant recipients hear more accurately. A schematic of a cochlear implant depicting both of the potential sites for the speech processor is shown in Figure 34.

3.2   Cochlear Implant Components and Surgery

As can be seen in Figure 35, the electronic code is sent to the transmitter coil (Figure 35, point 3) that is held above the implant (Figure 35, point 4) by a magnet. The transmitter coil uses an FM radio signal to transmit the signal through the intact skin into the implant. The implant package decodes the signal and sends a pattern of very rapid, small electrical pulses to

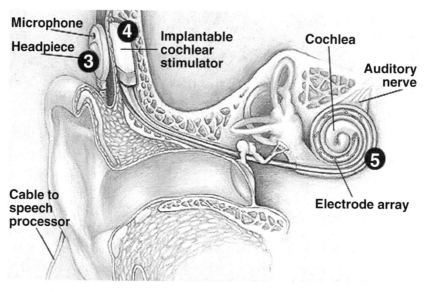

FIGURE 35: Cochlear implant components (see text for details).

the electrodes (Figure 35, point 5) in the cochlea, which stimulate the auditory nerves directly. Different parts of the nerve are stimulated according to the pitch of the sound waves. In response, the auditory nerve carries out its natural function and conducts nerve impulses to the brain. The brain receives the nerve impulses and interprets them as sound. This whole process takes place within a few milliseconds, corresponding to the time delay in the normally functioning ear.

The circuit is attached to a bundle of tiny wires that are inserted into the cochlea. At the end of the wires are as many as 24 electrodes that cover a distance of 25 mm along the length of the cochlea. Stimulation of each electrode usually causes a different pitch perception. A close-up view of the headpiece is shown in Figure 36, which indicates that the headpiece is often covered by the patient's hair and thus it is not normally evident. This is done on purpose to avoid the stigma associated with this relatively new technology for a profoundly deaf individual. The reasons for such a stigma will be discussed below.

Sounds are picked up by the small, directional microphone located in the headset of the ear. The microphone picks up all sound from the environment whether it is speech, environmental sounds, or music. Unlike a hearing aid, a cochlear implant is routinely used for the profoundly deaf individual. Although processing of speech is important, the transmission of other sounds is often as important, unlike a hearing aid where speech amplification is often the primary goal. Many profoundly deaf individuals may have never heard music or environmental sounds if deaf from birth. Thus the overall processing of sound with a cochlear implant has a greater impact

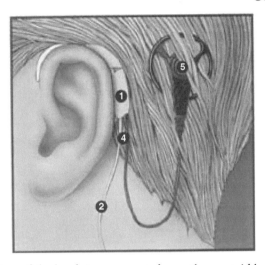

FIGURE 36: Close-up view of the headpiece connected near the mastoid bone.

and a broader goal than that of a hearing aid. An excellent review of the basics of cochlear implants has been published by Loizou (1998).

Sound received by the microphone must be processed to determine how the electrodes should be activated. The simplest way of sound processing would be to divide the auditory signal by the number of electrodes in the implanted portion of the device and apply appropriate voltage(s) to the appropriate electrode(s). In actuality, sophisticated processing algorithms are used, since applying voltage to each of the electrodes at the same time would cause currents to flow between the electrodes, which would stimulate multiple nerves and thus produce multiple (unwanted) frequencies in the "heard" sound. Modern sound waveform processing strategies often utilize bandpass filters to divide the signal into different frequency bands. Algorithms are employed to select a number of the strongest outputs from the filters, which often emphasize transmission of the time-domain aspects of uttered speech. Specialized feature extraction strategies use vowels and formants to evaluate each spoken utterance, the latter being an expanded set of vowel-like sounds. Fricative sounds are an expanded set of consonants and are utilized to fine tune the vowel and formant coding to allow proper microprocessor recognition of a spoken word or utterance. Normally, it is the time-varying aspects of the spectral content of speech that are employed in such coding. An example of this time-varying spectral content is shown in Figure 4. Figure 37 depicts speech coding strategies and elements of a cochlear implant including bandpass filters, envelope detectors, pulse generators, and culminating in the stimulating electrodes for a four-channel system (Loizou, 1998). The waveforms at each stage of this speech coding process are also displayed in Figure 37. The sound is processed through a set of four bandpass filters that divide the acoustic waveform into the four channels. Current

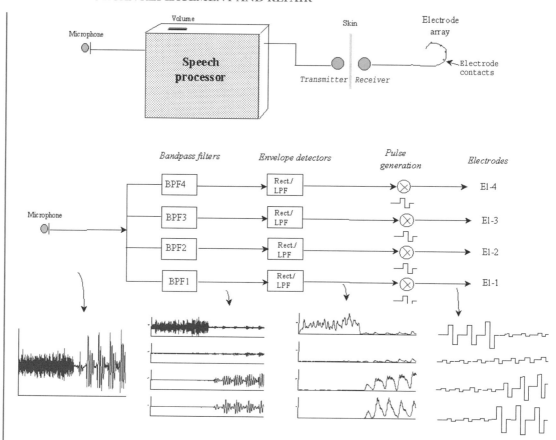

FIGURE 37: Stages of the speech processor in a typical cochlear implant and the associated processing waveforms at each stage for a four-channel device (Loizou, 1998).

pulses are generated with amplitudes proportional to the energy of each channel, and transmitted to the four electrodes through a radio frequency link. Channel 1 represents the lowest of the frequencies and channel 4 the highest. The amplitudes of each of the current pulses delivered to each of the electrodes reflect the spectral content of the input signal, with each appropriate electrode receiving excitation relative to the other electrodes. As an example, if the speech signal mostly contains high-frequency information, then the pulse amplitude of the fourth channel will be larger relative to the pulse amplitudes of the other channels. Similarly, if the speech signal mostly contains low-frequency information, then the pulse amplitude of the first channel will be larger relative to the amplitudes of the higher channels. The electrodes are therefore stimulated according to the energy level of each frequency channel, in a manner similar to that one in the natural cochlea.

Fundamental elements of a cochlear implant include electrode design (number of electrodes and electrode configuration), method of stimulation (analog or digital), as well as the method of feature extraction signal processing, as noted above. The design of electrodes for a cochlear implant has been the focus of research for over many years since the inception of the device (Clark *et al.*, 1983; Hochmair-Desoyer *et al.*, 1983). Early designs incorporated single electrodes in order to test the concept of a cochlear implant, such as that by Tyler (1998), with the number of electrodes employed rising with advances in the science and technology associated with cochlear implants. Modern versions employ 16, 24, or 32 electrodes. Issues associated with electrodes include electrode placement, number of electrodes, spacing of contacts, and orientation of electrodes with respect to the excitable tissue. A larger number of electrodes would provide better reproduction of sound to the brain, while a smaller number of electrodes would provide a more artificially sounding reproduction. However, more electrodes might introduce cross talk within the cochlea as closely spaced electrodes might produce excitation in neighboring tissue (below another electrode), even if that neighboring electrode did not fire. One issue is the sliding of the electrode set along the cochlea to rest at the appropriate place, with, as an example, the electrode representing the band 800–900 Hz being at the proper place in the cochlea where that would represent the same frequencies. This is both a surgery issue and a signal processing issue. It is vital that the electrodes are placed in close proximity with corresponding auditory neurons that lie along the length of the cochlea, or the entire sound spectrum will be shifted in frequency as "heard" by the brain. In most cases, the electrode arrays can be inserted in the scala tympani to depths of 22–30 mm within the cochlea.

The method of frequency encoding and electrode stimulation is constrained by the number of surviving auditory neurons that can be stimulated in the cochlea. It would be ideal to have surviving auditory neurons lying along the length of the cochlea, which would support a good frequency representation through the use of multiple electrodes, with each stimulating a different site in the cochlea as is done in a normal human ear. If the number of surviving auditory neurons is restricted to a small area in the cochlea, then the number of electrodes and frequency pattern is limited by pathophysiology. In that situation, only a few electrodes are needed near the surviving area of the cochlea. As such, it is important to have electrode design and numbers along with firing pattern and stimulation linked to the pathophysiology of the cochlea. It is pointless to use electrodes where there is no chance of providing neural response. Therefore, using a large number of electrodes will not necessarily result in better performance, because frequency coding and electrode stimulation are constrained by the number of surviving auditory neurons that can be stimulated.

In addition, electrode stimulation is constrained by the spread of excitation caused by standard electrical stimulation of tissue. When electric current is injected into tissue such as the cochlea, it tends to spread out symmetrically from the source. This is the basis for the

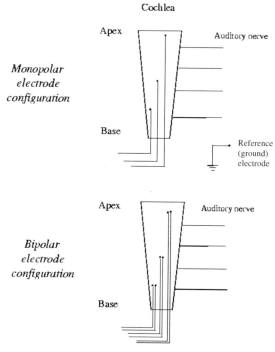

FIGURE 38: Monopolar and bipolar configurations of cochlear implant electrodes (Loizou, 1998).

electrocardiogram, as an example. As a result, the current stimulus in the cochlea does not stimulate just a single auditory neuron, but several. Such a spread in excitation is more prominent in a monopolar type of electrode configuration. In such a configuration, the active electrode is located far from a reference electrode, which acts as a ground for all electrodes in a multiple electrode configuration. The spread of excitation can be controlled to a degree by using a bipolar electrode configuration. In the bipolar configuration, the active and the reference (ground) electrodes are placed in close proximity to each other. Various studies have shown that bipolar electrodes produce a more localized stimulation than monopolar electrodes, such as Merzenich and White (1977) and van den Honert and Stypulkowski (1987), among others. Figure 38 depicts both types of electrodes.

There are generally two types of stimulation depending on how information is presented to the electrodes: analog or digital. In analog stimulation, an electrical analog of the acoustic waveform itself is presented to the electrode. As an example, in multichannel implants, the acoustic waveform is bandpass filtered, and the filtered waveforms are presented to the appropriate electrodes simultaneously in analog form. An advantage of analog signal processing is that the electronics are usually simpler and cheaper. A disadvantage of analog signal processing is that the simultaneous stimulation may result in cross talk. In digital signal processing, the information is delivered to the electrodes using a set of pulses. In some approaches, the timing and amplitude of these pulses are obtained from the envelopes of the filtered waveforms, as is shown in Figure 37. The advantage of this type of signal processing is that the pulses can be delivered in a nonoverlapping approach, thereby minimizing cross talk. The rate at which these pulses are delivered to the electrodes has been found to affect speech recognition performance (Wilson et al., 1995). High pulse rates tend to yield better performance than low pulse rates. Pulse rates as low as 100 pulses per second and as high as 2500 pulses per second have been used. Wilson et al. reported that some patients obtained a maximum performance with a pulse rate of 833 pulses per second and a pulse duration of 33 ms, while other patients obtained significant increases in performance as the pulse rate increased from 833 to 1365 pulses per second, and from 1365 to 2525 pulses per second, again using 33-ms pulses (Wilson et al., 1995). It is apparent that an optimal pulse rate depends on a particular patient, but that the pulse width is dependent on the tissue sustaining the electrical excitation. A digital representation of the speech signal often results in a more accurate representation, but the technique may be more expensive. Modern cochlear implants employ digital signal processing techniques for multiple electrode configurations. Another benefit of the digital technique is that the digital pulse stimulation order (to the various electrodes) can be varied to minimize possible cross talk between channels. Electrodes can be sequentially excited (in a rapid fashion), randomly excited, or excited in groups with the spacing far enough apart to minimize cross talk. As with the pulse rate, the optimal electrode firing pattern may vary from patient to patient.

Currently, some implant devices employ monopolar electrodes, while some devices employ bipolar electrodes, and there are other devices that employ both types of electrodes. Loizou reported that the Symbion cochlear implant used six electrodes spaced 4 mm apart in their bipolar configuration with only the four most apical electrodes used in the monopolar configuration (Loizou, 1998). The Nucleus cochlear implant used 22 electrodes spaced 0.75 mm apart for the monopolar configuration with electrodes that were 1.5 mm apart used as bipolar pairs. The Clarion cochlear implant used both monopolar and bipolar configurations. Eight electrodes spaced 2 mm apart were used. The Med-El cochlear implant used eight electrodes spaced 2.8 mm apart in a monopolar configuration. Further details regarding the electronics and signal processing elements in various cochlear implants have been reported by Loizou (1998).

3.3   Current Devices and Cochlear Implant Companies

In 2005, the top three cochlear implant devices were manufactured by Cochlear Corporation (Australia), Advanced Bionics (United States), and MED-EL (Austria). These are similar devices. Each manufacturer has adapted some of the successful innovations of the other companies to their own devices. There is no clear-cut consensus that any one of these implants is superior to the others. Users of all three devices display a wide range of performance after implantation.

Of the three companies, Cochlear Corporation has the longest history in cochlear implantation and the high reliability of their devices has been demonstrated over three decades, as the cochlear implant is generally known to have originated in Australia. Advanced Bionics has been noted to have more advanced electronics, resulting in the more advanced speech processing capabilities. MED-EL has the longest electrode array of the three companies, enabling insertion into the deepest part of the cochlea.

Since the devices have a similar range of outcomes, other criteria are often considered when choosing a cochlear implant: usability of external components, cosmetic factors, battery life, reliability of the internal and external components, customer service from the manufacturer, the familiarity of the user's surgeon and audiologist with the particular device, and anatomical concerns.

There is a great variability in the speech recognition performance of cochlear implants as reported by patients, technicians, and physicians. For a given type of implant, auditory performance (in terms of word or utterance recognition) may vary from zero to nearly 100% correct. The factors responsible for successful word/utterance recognition have been the focus of research for many years. Some of the factors that have been found to affect auditory performance are listed below:

1. *Outside factors*: These include duration of deafness, age at implantation, and residual hearing, as well as issues that are beyond our control. There is a huge variance in how well people do with implants, largely due to the wide degree of differences of these factors. Those with residual hearing respond better to cochlear implants and require less training. Those patients who are deaf from birth require more training and have a checkered success rate. Younger patients do better than older patients who have been deaf for longer periods. Very young patients, who have not yet formulated language, respond better. A discussion of cochlear implants in the very young patients is noted below as are factors associated with the deaf community response to cochlear implant usage.

2. *Attitude and determination*: The amount of work and practice that the patient puts into making it work is very important to making the most of the implant. Cochlear implant recipients normally receive speech and/or auditory therapy to make the most of their implants. The implant is not a miracle cure for deafness, since the number of channels is far less than the number of frequency excitations in the natural ear. Many profoundly deaf (or deaf from birth) patients find it frustrating and discouraging to undergo the long and difficult process of learning to hear with the implant.

3. *Skill of audiologist and surgeon*: A good surgeon and audiologist can help ensure that recipients get the most that the implant has to offer. One should determine how much

experience the audiologist and the surgeon have with implants, and get a sense for how dedicated they may be to help you in the years to come with any challenges that arise.

4. *Type of device*: Newer generations of cochlear implants have improved the chances that people will hear well with them, with digital signal processing techniques, improved electrode design, improved support electronics, and improved surgical techniques.

3.4 Issues Associated with Implant Use

As was noted above, there have been issues raised regarding the use of cochlear implants, which has limited its implantation in a large number of patients. For the elderly patient, the risk of surgery in the older patient must be weighed against the improvement in quality of life. As the devices improve, particularly the sound processor hardware and software, the benefit is judged to be worth the surgical risk particularly for the newly deaf elderly patient (Waltzman *et al.*, 1993).

The use of cochlear implants is objected to by some, particularly in the deaf signing community (people who use sign language to communicate). Many members do not view deafness as a disability that has to be "fixed" by cochlear implants, but rather just a different way of living. Cochlear implants for children work best when implanted at a young age, when the brain is still learning to interpret sound, and hence are implanted before the recipients can decide for themselves. There is a debate about the age that would be the best or safest for children to receive cochlear implants. However, as cochlear implants have improved and proven themselves to work well, the objections are getting less tenable. Proponents of cochlear implants believe that, since mammals are meant to have a hearing sense, deafness is a disability to be corrected. To them, objecting to a cochlear implant is akin to objecting to a heart transplant, prosthetic arm, or glasses.

Opponents of cochlear implants often compare implantation to cultural genocide, as the common ground for the signing community is deafness, either through personal experience or by personal relationships with others who have hearing impairments. Opponents of implantation tend to object not to the medical benefits of implantation, but to the perceived loss of the sense of community that has defined deaf history. Statistics show that the majority of deaf infants are born into hearing families. For such children becoming a member of the deaf community is not automatic, nor is it initially a decision for them to make; rather it is a choice made by their parents, in what they perceive to be the children's best interests, as they understand them. In 2000, an academy-award-nominated film *Sound and Fury* depicted this cultural divide. In addition, the implantation of cochlear implants in very young children has become a topic of much consternation in the deaf community and has spawned Web sites such as www.cochlearwar.com and www.ragged-edge-mag.com, both have numerous citations and editorials opposing the implantation of cochlear implants in small children. Of all artificial organs and all sensory assist or replacement devices, the use of cochlear implants has created the largest amount of controversy and vitriol. This is primarily due to the strongly cohesive nature of the deaf community that does not feel that deafness is a disability like paraplegia or blindness, and thus does not necessitate the use of a cochlear implant. Although there is less controversy with the implantation of cochlear implants in adults, there is still not widespread support for such implantation, which has reduced the number of individuals seeking an alternative to profound deafness. Opponents of cochlear implants cite the prevalence of acquired meningitis as a result of implantation surgery. However, data from the FDA and the Center for Disease Control (CDC) indicates that the prevalence of meningitis for the general population varies from 1 in 10 000–15 000 and for those with cochlear implants, it is 1 in 9000–13 000. There is thus statistically no significant difference. However, the data for the general population utilize a far greater sample size than that for the cochlear implant patient data.

3.5   History of the Cochlear Implant

Professor Graeme Clark of the University of Melbourne is generally credited as the creator and developer of the world's first multichannel implant and is considered by many to be the father of the cochlear implant. In 1967, Professor Clark began researching the possibilities of an electronic implantable hearing device. His endeavor would eventually lead to the creation of the implantable "bionic ear." Inspired by his close relationship with his father, who had been deaf throughout his life, Professor Clark's goal was to find a way to improve hearing, and the quality of life for people who are deaf. Professor Clark has been the Research Leader at the Department of Otolaryngology at the University of Melbourne since 1970. His first cochlear implant surgery took place in 1978. In 1979, Nucleus, a group of medical equipment manufacturers, became interested in the commercial potential of Professor Clark's work at the University of Melbourne. Two years later, the University of Melbourne, the Australian Government, and Nucleus set out to develop a commercially viable cochlear implant, and to bring it to market through a worldwide clinical trial. Cochlear Ltd. was established in 1982, as a corporate entity in order to continue commercial operations. Professor Clark is the Director of The Bionic Ear Institute in Melbourne (Australia), which was established in 1984 to support and undertake research to better understand deafness and improve the "bionic ear."

The basis for the cochlear implant is predicated on the spatial frequency representation of the cochlea, which was first elucidated by the pioneering work of Georg von Bekesy in the 1950s, which showed that the basilar membrane in the inner ear is responsible for analyzing the input signal into different frequencies (Hachmeister, 2003; Shampo and Kyle, 1993). Different frequencies cause maximum vibration amplitude at different points along the basilar membrane. A famous diagram of this spatial frequency representation is shown in Figure 39.

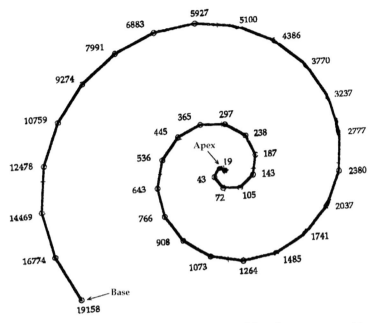

FIGURE 39: Spatial variation by frequency in the human cochlea, first represented by von Bekesley.

3.6   Regulatory Issues for Cochlear Implants and Recent Research

For a relatively new technology such as a cochlear implant with few implants compared to the entire deaf population, the issues regarding regulatory controls for cochlear implants are of paramount importance. This is very evident when one looks at the FDA Web site at www.fda.gov and then to its medical devices section, where cochlear implants are one of the key topics on the right (as of January 2006). The FDA has a primer on the topic complete with movies. From 1984 through 2001, there were seven different applications approved for cochlear implants including the following:

Cochlear Corp	1984	3M Brand
Cochlear Americas	1985	Nucleus Mutichannel
Cochlear Americas	1990	Nucleus 22 Channel
Advanced Bionics	1996	Clarion Multistrategy
Advanced Bionics	1997	Clarion Multistrategy
Cochlear Americas	1998	Nucleus 24
Med-El Corp	2001	Combi 40+

The Web site listing these approved applications is at http://www.accessdata.fda.gov/scripts/cdrh/devicesatfda/index.cfm, which also notes any warning letters, MedWatch reports, or enforcement reports for each device. There have been no approved devices by the FDA since 2001.

In 2004, Advanced Bionics issued a voluntary recall of its Clarion cochlear implants because there were noted moisture problems prior to implantation for some of their devices. This report is reported at http://www.fda.gov/bbs/topics/news/2004/NEW01119.html. The FDA also issued a report on the risk of bacterial meningitis in small children receiving cochlear implants. This report is at http://www.fda.gov/cdrh/safety/cochlear.html.

As with any technology that has received few approved applications over 20 years including none over the last 5 years, there are still significant questions to answer. Is the use of the device hampered by the nonsupport of the deaf community or due to the checkered success rate for profoundly deaf patients (from birth) who have no basis to understand spoken language as processed by a cochlear implant? The FDA Web site that analyzes risks and benefits ishttp://www.fda.gov/cdrh/cochlear/RiskBenefit.html.

There have been numerous published reports regarding clinical and empirical research on cochlear implants. These can generally be categorized as reports regarding the use of cochlear implants in children, the use of cochlear implants in the elderly, a comparison of cochlear implants versus hearing aids, and studies regarding design issues, such as electrode configuration and signal processing techniques. Recent studies regarding the use of cochlear implants in small children include those by Beadle *et al.* (2005), Berg *et al.* (2005), Biernath *et al.* (2006), Flipsen and Colvard (2005), Gordon *et al.* (2005), Henkin *et al.* (2005), Higgins *et al.* (2005), Hyde and Power (2005), and Tomblin *et al.* (2005), among others.

Studies regarding the use of cochlear implants in the elderly include those by Haensel *et al.* (2005) and Hay-McCutcheon *et al.* (2005), among others, while the studies regarding design considerations include Fabry (2005), Fu *et al.* (2005), Hong and Rubinstein (2006), Laneau *et al.* (2005), Munson and Nelson (2005), Stickney *et al.* (2005a), Stickney *et al.* (2005b), Vermeire *et al.* (2006), and Xu *et al.* (2005), among others.

Despite the relatively small numbers of FDA approved devices, the number of worldwide implants is growing, and the number of clinical studies and published reports number in the thousands.

4   INTRAOCULAR LENS

4.1   Anatomy of the Eye

An IOL is used to replace a damaged lens in the eye, most normally due to a cataract, which results in cloudy vision. The human eye, shown in Figure 40, consists of the multilayer outer covering (the cornea), the iris and associated pupil opening, the lens inside a lens sac, and the retina. There are fluid layers between the cornea and the lens and again between the lens and the

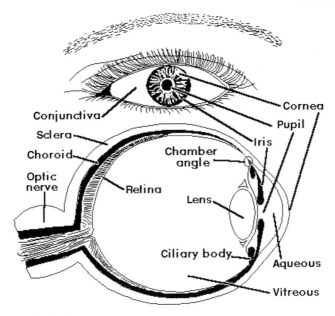

FIGURE 40: Anatomy of the human eye.

retina. The retina ends with the optic nerve. The cornea provides 75% of the overall focusing power of the eye with the lens proving the other 25%.

However, it is the ability of the natural lens to *accommodate*, to change focal length, and the fine tuning of accurate sight that is key to visual field acuity. The human lens is a gelatinous substance, which is flexible. This is important as the sac in which it resides is connected on either end by ciliary muscles. These muscles can rapidly move in and out. As your visual attention moves closer or farther away, the ciliary muscles (sometimes called the ciliary body) attached to your lens sac move in and out, which changes the shape of the lens to make it more or less concave and provide variable focus. This process is called accommodation. When one looks directly at one's feet and then suddenly attempts to focus 50 yards away, the focus remains true and one's visual field remains perfectly clear. This could not happen without the lens being able to change its shape as the lens sac is pulled by the attached muscles. The continuing change in the shape of the lens allows the focal point of vision to fall on a section of the retina known as the *fovea*. A more detailed view of the human eye showing the fovea on the retina is shown in Figure 41.

4.2   Cataracts and Their Determination

A cataract is a clouding of the human lens with the resulting vision becoming opaque in that eye. Often, if only one eye is affected, it may be possible that an individual does not realize that a

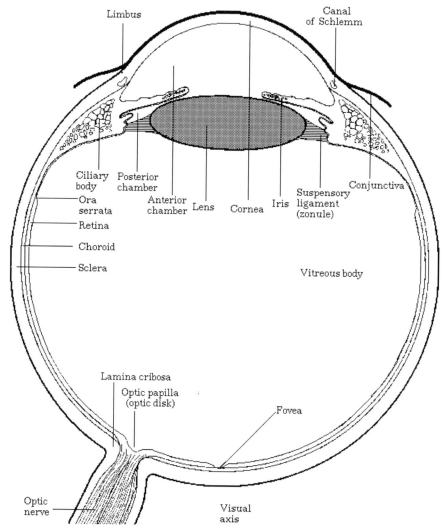

FIGURE 41: Anatomy of the human eye showing the fovea portion of the retina.

cataract is evident. Over 40 million people in the United States suffer from cataracts. A cataract can result from many sources, including age, diabetes, glaucoma, blunt injury, excessive infrared light exposure, excessive steroids, and genetic effects. The natural lens, which is crystalline in nature, becomes opaque and less flexible with an advanced cataract. This results in the lens becoming less able to accommodate by changing shape. However, it is the opaque vision effect of a cataract that is often more telling in both the diagnosis and the patient complaint. A routine eye examination with a slit lamp would indicate a cataract, since the eye would show whiteness in the lens indicating cloudy vision. A typical slit lamp examination is shown in Figure 42.

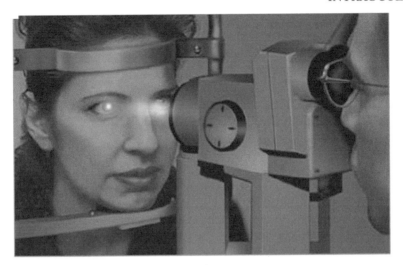

FIGURE 42: Typical slit lamp examination to examine the interior of the eye.

Figure 43 shows both a normal eye (front view) and an eye with a cloudy lens. It is obviously easy to see this in a routine eye examination. The resulting visual field is different for a normal lens and a cloudy lens, as is shown in Figure 44. The opaque nature of the visual field is easier to notice by the patient in sunlight and is less easy to notice in artificial (indoor) light.

Once a cataract has been determined to exist, it is necessary to evaluate the location of the lens, its distance from the cornea and from the retina, the thickness of the lens, and the size of the sac. This will eventually allow the clinician to determine the correct power and size of an artificial lens, the *intraocular lens*. This geometric and anatomical information of the eye requires detailed measurements within the fluid-filled eye with a resolution that is on the order of 0.1 mm. The use of ultrasound at a high frequency is perfect for this determination for several reasons: 1) ultrasound transmits well within a liquid; 2) at a 10–15-MHz emission frequency, the resolution is on the order of the required 0.1 mm; and 3) the attenuation factor for ultrasound

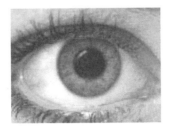 

FIGURE 43: Views of human lens: normal lens on the left and cloudy lens on the right.

Normal vision   Vision through
a cataract

FIGURE 44: Visual field for a normal lens on the left and a cloudy lens on the right.

is lessened due to the small depth of the eye. A typical A-scan ultrasound examination of the eye is shown in Figure 45.

The ultrasound probe is a small 3-mm rod attached to a spring-loaded micrometer device, sometimes called a Goldman tonometer holder. The clinician slowly rotates a knob and the probe slowly nears the patient's eye. When the probe contacts the cornea and meets resistance, it automatically springs backward away from the patient. This actually gives the sufficient time to make the appropriate measurement, since the 10-MHz emission frequency is actually pulsed 1000 times per second. As a result, each 1-ms pulse width contains 10 000 emitted pulses at the 10-MHz frequency (each emitted wave being 0.1 μs). Thus, in as little as 1 ms of contact

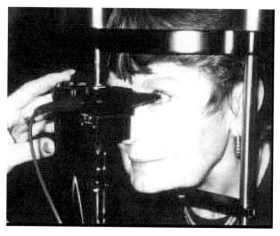

FIGURE 45: A scanner for the eye.

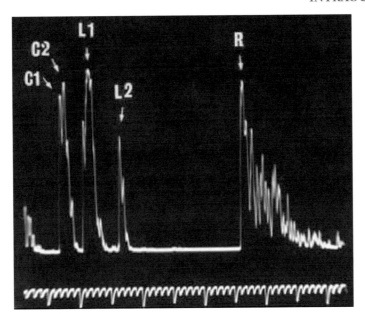

FIGURE 46: A-scan ultrasound pulse-echo pattern in the eye.

with the cornea, the A-scan probe would have emitted 10 000 pulses with the returning echoes processed for turnaround time. This turnaround time duration is equivalent to the distance from the probe (and thus the cornea outer edge), since a constant speed of sound is assumed. A typical A-scan ultrasound pulse-echo pattern is shown in Figure 46.

The C1 and C2 echoes in Figure 46 refer to the anterior and posterior surfaces of the cornea, while the L1 and L2 echoes refer to the anterior and posterior surfaces of the lens, and the R echo refers to the anterior surface of the retina. The various distances from the cornea to the lens and from the lens to the retina dictate the power of whatever replacement lens is required. Any variation in measurements of these distances can affect the final power of the implanted lens, and this depends on the axial length (depth) of the eye. In an average 23.5-mm eye, a 0.1-mm difference in axial length measurement (of the eye) affects the final postoperative refraction by 0.25 D (D = diopter, the standard optical power for lenses). In a long 26.0-mm eye, a 0.1-mm difference in the AL measurement affects the final postoperative refraction by only 0.20 D. In the short 21.0-mm eye, a 0.1-mm difference in the axial length measurement affects the final postoperative refraction by 0.31 D.

4.3   The IOL and Implantation Surgery

An IOL is a replacement device for a cloudy natural lens. Because of the small size of the natural lens and the lens sac, along with a need to have very small incisions for removal of the old lens and insertion of the replacement lens, there is a size restriction on the artificial lens and any device to hold this lens. Furthermore, as this is a device that will be in close contact with liquids and tissues inside the body, this device must be biocompatible and nonreactive. Obviously, this device must have the correct optical power as determined by the A-scan ultrasonic examination of the eye. As will be described below, the standard IOL is pliable and can be folded in half during insertion, so as to limit the size of the incision.

However, first things first. The cloudy natural lens must somehow be removed in order to have space for the replacement IOL. As the natural lens is a crystalline, gelatinous material, we can employ various technologies to remove this lens from the eye. The most commonly employed technique is to use high-frequency, therapeutic level ultrasound, often called *phacoemulsification*. Unlike the A-scan ultrasound used to measure eye distances, the ultrasound used to emulsify the natural lens is not a pulse-echo device, but rather a continuous wave device. The pulse-echo device is actually "on" for 1% of the time with the other 99% in the echo mode to detect the time course of returning echoes. In the continuous wave mode, particularly with higher amplitudes than the pulse-echo mode, the tissues or fluids can vibrate and even heat up. Therapeutic ultrasound is sometimes employed to ease aching muscles by heating the tissue and increasing blood flow to the region. In phacoemulsification, the gelatinous lens is liquefied (emulsified) and then aspirated out, leaving an empty sac behind.

The cataract surgical procedure consists of the following steps:

1. An outer incision to the edge of the cornea
2. An inner incision in the lens sac
3. Phacoemulsification of the natural lens
4. Aspiration of the liquefied natural lens
5. Insertion of the folded IOL
6. Suture the lens sac with dissolvable sutures
7. Suture the outer incision with nondissolvable sutures

Both incisions are approximately 3 mm long. Figure 47 depicts the miniature scalpel used to create the microincisions. The surgeon creates an opening in the lens capsule, which is a microthin membrane surrounding the cataract. This procedure, called capsulorhexus, requires extraordinary precision since the capsule is only about four-thousandths of a millimeter thick.

A typical phacoemulsification probe is shown in Figure 48 with the ultrasound waves seen emanating within the lens.

Figure 49 shows the aspiration portion of the process once the lens has been fully emulsified (liquefied). The aspiration probe is moved around the sac to completely aspirate all loose materials within the sac, as is shown in Figure 49.

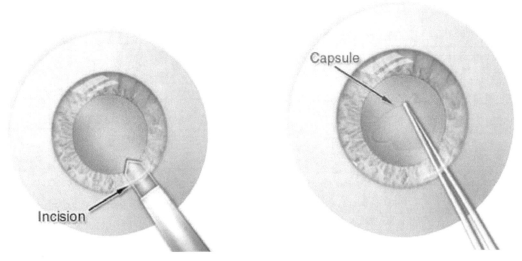

FIGURE 47: Incisions in cataract surgery: left—on the side of the cornea with a miniature scalpel and right—on the side of the lens sac.

Once the original cloudy lens has been removed, it is then possible to insert the IOL. A typical IOL is shown in Figure 50 and consists of a biconvex, clear lens with "wings" to hold the lens in place within the lens sac/capsule.

The springlike wings pop out when the IOL is unfolded during insertion. A diagram with the IOL in place within the eye as compared to the natural lens is shown in Figure 51.

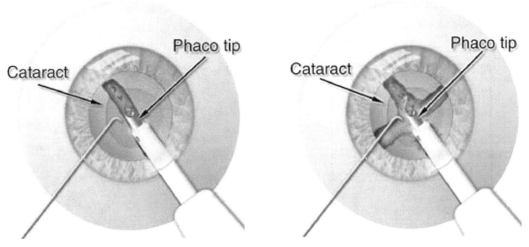

FIGURE 48: Phacoemulsification showing the sound waves with breakup of lens to left.

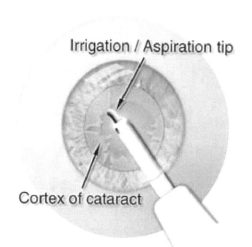

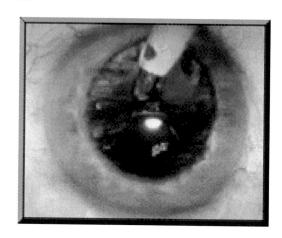

FIGURE 49: Irrigation and aspiration of the emulsified lens. Aspiration probe is moved around the lens sac to aspirate all particles and lens pieces.

A typical IOL is a clear lens made of an acrylic polymer with 25–30% water content, so that it will not absorb water when inserted into the fluid-filled environment of the human eye (aqueous and vitreous regions as noted in Figures 40 and 41). An IOL has various optical powers based upon patient needs. It is typically 6–8 mm in diameter and is available in a power range between 10 and 30 D in 0.5-D increments. Other IOL's are made of silicone, poly(methyl methacrolate) (PMMA), or a clear polymer known as perspex. Many IOL designs have ultraviolet light blockers incorporated into the lens material.

As was stated earlier, it is necessary to fold the IOL during insertion, as the 6–8-mm diameter is larger than the incision width of 3–4 mm. Thus, specialized tools are used to fold and

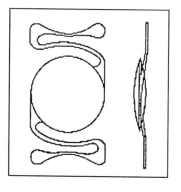

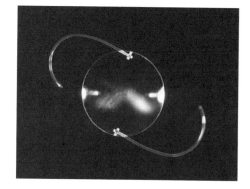

FIGURE 50: Typical IOL with wings.

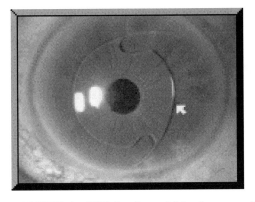

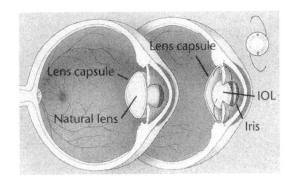

FIGURE 51: IOL in place within the eye and lens sac/capsule.

hold the IOL as well as release it once it is placed within the lens sac/capsule. A forceps design is offered by various IOL manufacturers which folds and holds the lens while it is inserted into the capsule. Typical forceps designs are shown in Figure 52.

Once the IOL is placed within the lens capsule, the IOL is then released and it pops open within the capsule with the spring-loaded wings extending outward. Other forms of IOL holders include injection-type devices with push button or screw-turned hold and release mechanisms, as is shown in Figure 53.

Following implantation, the slit at the edge of the lens sac/capsule is often sutured with dissolvable sutures. The outer slit at the edge of the cornea is closed with nondissolvable sutures. The patient is given strong wraparound sunglasses, as there will be a strong light sensitivity for

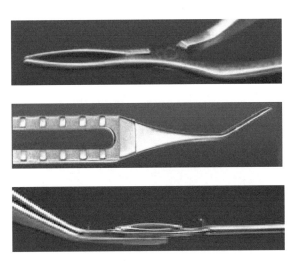

FIGURE 52: IOL forceps designs that hold the folded IOL for insertion into the lens sac.

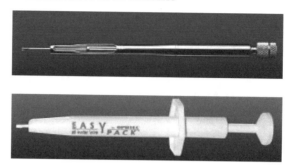

FIGURE 53: IOL injection style holders with push button or screw release.

a few days. In addition, the patient is given a plastic disk that must be taped over the affected eye at night. This is to protect the eye from accidental rubbing during the night. Normally, the patient is seen by the ophthalmic surgeon the day after surgery, in order to check the incision, the healing process, and to determine if there is any swelling or infection. Approximately 1 week after surgery, the patient returns to have the outer sutures removed. The ophthalmic surgeon carefully snips the ends of the sutures and uses a small tweezers to pull the sutures out. The patient can feel the light pulling of the sutures, but rarely feels any pain.

On some occasions, the lens sac/capsule may become cloudy at some point following the surgery as is shown in Figure 54. This is normally remedied by means of an out-patient visit to the surgeon's office. An excimer laser is used to burn small holes in the sac at various points. The patient is placed in a head mount device and the affected eye is held open. The patient sees red flashes and hears "pops," but does not feel anything during the process. The holes in the sac supply sufficient light such that the clear vision is restored. The sac is otherwise still intact.

At times, a patient may experience a "floater." Floaters, or *muscae volitantes* (Latin—"flying flies"), are characterized by shadowlike shapes that appear singly or together with several others in one's field of vision. They can take the form of spots, threads, or fragments of cobweblike

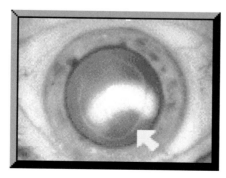

FIGURE 54: Lens sac becomes cloudy some time after surgery.

shapes that float slowly before one's eyes. Floaters are suspended in the vitreous humor, the thick fluid or gel that fills the eye. Thus, they generally follow the rapid motions of the eye while drifting slowly within the fluid. When they are first noticed, the natural reaction is to attempt to look directly at them. However, attempts to shift the gaze toward them are frustrating, because the floaters follow the motion of the eye, and remain to the side of the direction of gaze. Although the blood vessels of the eye also obstruct light, they are invisible under normal circumstances (and thus not annoying) because they are fixed in the location relative to the retina, and the brain "tunes out" stabilized images. This does not occur with floaters and they remain visible, and, in some cases, when large and numerous, are very irritating. Despite the name "floaters," many of these specks have a tendency to sink toward the bottom of the eyeball, in whichever way the eyeball is oriented. Floaters are not uncommon, although they rarely cause problems for those who have them. Floaters can be a nuisance and a distraction to those who suffer from severe cases, as the spots seem to drift through the field of vision. Normally, there is no treatment indicated, particularly for mild cases, as the floater will eventually sink toward the bottom of the vitreous body and out of the field of vision. Surgical or laser treatments have been employed for severe cases with mixed results.

4.4   History of the IOL and Current Research

Nicholas Harold Lloyd Ridley, most commonly known as Harold Ridley, was a British ophthalmologist who pioneered IOL surgery for cataract patients. He was famous as a Surgeon at Moorfields Eye Hospital and St. Thomas' Hospital in London, specializing in ophthalmology. While working with Royal Air Force casualties during World War II, he noticed that when splinters of perspex from aircraft cockpit canopies became lodged in the eyes of wounded pilots, they did not trigger rejection, leading him to propose the use of artificial lenses in the eye to correct cases of cataracts. He had a lens manufactured using an identical plastic, and on November 29, 1949, at St. Thomas' Hospital, he achieved the first implant of an IOL, although it was not until 1950 that he implanted an artificial lens permanently in an eye. He went on to develop widespread programs for cataract surgery with intraocular implants and pioneered this treatment in the face of initially strong opposition from the medical community. He continually refined the technique; until by the late 1960s, with his pupil Peter Choyce, he eventually achieved widespread support for the technique. The IOL was finally approved for use in the United States by the FDA in 1981. IOL implantation and cataract surgery is now a common type of eye surgery.

Although the most common form of an IOL is for the treatment of a cataract, there is now another form of IOL used for the treatment of myopia. The alternative IOL is placed in the anterior chamber (with respect to the lens capsule) and is primarily used to treat nearsightedness, as opposed to a cataract. In this fashion, it is similar to that of glasses or contact lenses, but is permanently placed within the eye, as is done for more standard IOLs. This new IOL is the latest type to receive FDA approval as noted at http://www.fda.gov/bbs/topics/ANSWERS/2004/ANS01313.html. As befits a new style of IOL for a different purpose that was originally intended to treat cataracts, there have been numerous studies of late including those by Alio *et al.* (2005), Benedetti *et al.* (2005), Leccisotti and Fields (2005), Lifshitz *et al.* (2004), Lovisolo and Reinstein (2005), and Sridhar *et al.* (2005).

For traditional IOLs to treat cataracts, recent studies have focused on the materials used for an IOL as well as surgical and follow-up complications of various IOL designs. Examples of studies involving materials used for an IOL include a study of the use of PMMA by Frazer *et al.* (2005) as well as a comparison of PMMA, silicone and acrylics by Smith *et al.* (2005). Examples of studies involving surgical and postsurgical complications of an IOL include those by Cakmak *et al.* (2005), Collins *et al.* (2006), Martin and Sanders (2005), Oshika *et al.* (2005), Parsons *et al.* (2005), Tassignon *et al.* (2005), Tognetto *et al.* (2005), and Werner *et al.* (2006).

5   ARTIFICIAL AND REPLACEMENT CORNEA

The cornea is the outer coating of the human eye as can be visualized in Figures 40 and 41 and provides most of an eye's optical power. Together with the lens, the cornea refracts light and consequently helps the eye to focus. The cornea gives a larger contribution to the total refraction than the lens, but whereas the curvature of the lens can be adjusted to alter the focus based on accommodation, the curvature of the cornea is fixed. The cornea has sensitive nerve endings such that a touch of the cornea causes an involuntary reflex that closes the eyelid. As transparency is of prime importance, the cornea does not have blood vessels, but rather receives nutrients via diffusion from tears on the outside surface and the aqueous humor on the inside surface. The adult cornea has a diameter of approximately 12 mm (about one half inch) and a thickness of 0.5–0.7 mm in the center and 1.0–1.2 mm at the periphery. In humans, the refractive power of the cornea is approximately 45 D, which is approximately three-fourths of the eye's total refractive power. The cornea consists of five layers. From the outside to the inside they are as follows:

- *Corneal epithelium*: a thin epithelial layer of fast-growing and easily regenerated cells. Tears keep this layer moist.

- *Bowman's layer*: a tough layer that protects the corneal stroma. It consists of irregularly arranged collagen fibers.

- *Corneal stroma*: a thick, transparent middle layer responsible for most of the focusing power of the cornea. It consists of regularly arranged collagen fibers along with a few fibroblasts. If the stroma is damaged, for example, by injury or infection, it can lose its transparency, causing vision problems. The corneal stroma consists of approximately 200 layers of type I collagen fibrils. The ordering of the fibrils is responsible for the transparency of the tissue.

- *Posterior limiting membrane (also called Descemet's membrane)*: a thin acellular layer that serves as the modified basement membrane of the corneal endothelium.
- *Corneal endothelium*: a simple scaly style epithelium, an inner lining acting as a barrier to prevent water inside the eyeball from moving into and hydrating the cornea, which would lead to blurred vision.

The cornea is composed mostly of dense connective tissue, similar to the surrounding sclera. However, the collagen fibers are arranged in a parallel pattern, allowing light waves to constructively interfere, letting the light pass through relatively uninhibited. The cornea is innervated by the long posterior ciliary nerves.

Various refractive eye surgery techniques, such as LASIK surgery, can change the shape of the cornea in order to reduce the need for glasses or otherwise improve the refractive state of the eye. In the techniques used today, parts of the cornea are removed with lasers. If the corneal stroma has developed opaque patches known as leukomas, a cornea of a deceased donor can be transplanted. Because there are few blood vessels in the cornea, there are also few problems with rejection of the new cornea. Replacement corneas are normally obtained from eye banks, which store donor corneas. There are also synthetic corneas in development. Most are merely plastic inserts, but there are also some made of plastics that encourage the eye tissue to grow into the synthetic cornea, making it a full replacement. In addition tissue-based artificial corneas are now in development.

5.1   Corneal Transplant

A corneal transplant is the replacement of a damaged cornea with an undamaged one. Corneal transplant is also called keratoplasty. A corneal transplant is not the same thing as laser-assisted *in situ* keratomileusis; LASIK involves reshaping, not removal, of the cornea, in order to correct focusing problems. Although the cornea is tough, due to its vulnerable location, it is quite susceptible to damage from accidents, bacteria, and particles in the air. The well-organized arrangement of the collagen is required in order for the cornea to remain transparent. Corneal transplantation is undertaken for blurry vision resulting from a cornea that no longer enables light to pass through uninterrupted, due to cloudiness or scarring. Reasons for opaqueness of the cornea may be

- degenerative disease;
- dystrophies (inherited diseases in which impurities gradually accumulate in the cornea);

- infection (infection of the cornea can also occur as a complication of other infections); and

- trauma.

In the procedure, the central region of the cornea is cut out with a trephine (an instrument resembling a cookie cutter). The replacement cornea is then inserted and sutured into place using extremely fine thread (narrower than a human hair). The patient is awake during the operation, which is conducted under local anesthesia. The eye is held open by special device. Surgery requires 30–90 min. After the surgery, eye drops are used to aid healing. Nevertheless, recovery from keratoplasty occurs extremely slowly; complete healing takes months to years, and so the sutures remain in place for at least several months. Complications include the possibility of rejection of the foreign tissue graft or bleeding or infection risks of surgery. The use of large amount of steroids following the operation improves the success rate. Nevertheless, about one in five patients rejects the new cornea. Over 40 000 corneal transplants are performed in the United States every year. Medicare reimbursement for a corneal transplant in one eye was about $1200 in 1997. Figure 55 depicts a cornea before and after corneal transplant surgery.

Figure 56 depicts the sequence of the surgical procedure for corneal transplantation.

There are numerous regional and local eye banks dedicated to receiving donor corneal tissue with proper storage until needed by a patient with a cloudy cornea. One can contact local eye banks by looking in the local yellow pages under "eye bank" or by searching yahoo.com under "eye bank, state name (or city name)." There will be links on how to become a donor including downloading a donor card. Physicians involved with corneal transplantation usually know how to access the local eye bank for donor tissue. Although blood-typing is not a necessity (as the

Before   After

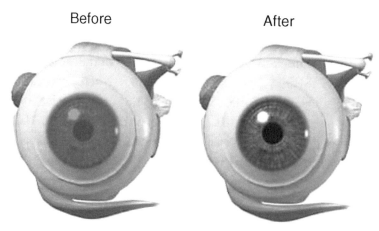

FIGURE 55: The cornea before and after transplant surgery. The cloudy cornea on the left is replaced with a clear transplant cornea on the right.

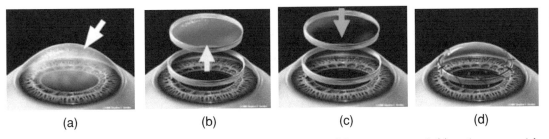

| (a) | (b) | (c) | (d) |

FIGURE 56: Corneal transplant surgery: (a) damaged cornea; (b) cornea removal; (c) replacement with donor cornea from eye bank or cadaver; and (d) suture replacement cornea in place.

corneal is avascular), it is often prudent to match blood type between donor and patient. The United Network for Organ Sharing, which is responsible for organ donation and transplantation for a variety of organs, does not handle eye tissue. This is done via eye banks at the local and regional level.

Replacement corneas that are either plastic or tissue based have been developed in recent years. An artificial (nontissue) cornea has been developed by AlphaCor produced by Argus Biomedical Corp. AlphaCor is a biocompatible, flexible, one-piece artificial cornea (kerato-prosthesis) designed to replace a scarred or diseased native cornea. It is technically listed as a *hydrogel*. AlphaCor is designed for use in patients who have had multiple failed corneal transplants or in those patients in whom a donor graft is likely to fail. AlphaCor is available in two versions, to suit those with a natural lens (phakic) or artificial lens (pseudophakic) and for those without a lens (aphakic). AlphaCor is a one-piece convex disk consisting of a central transparent optic and an outer skirt that is entirely manufactured from poly(2-hydroxyethyl methacrylate) or PHEMA. The outer skirt is an opaque, high water content, PHEMA sponge. The porosity of the sponge encourages biointegration with host tissue and thus promotes retention of the implanted device. The central optic core is a transparent PHEMA gel providing a refractive power similar to that of the human cornea. The junctional zone between the skirt and the central optic is the interpenetrating polymer network (IPN). This is a permanent bond formed at the molecular level and is designed to prevent the down growth of cells around the optic, which can lead to the formation of retroprosthetic membranes, one of the major complications historically associated with artificial corneas. The AlphaCor device is shown in Figure 57.

The implantation of AlphaCor is a unique two-stage procedure. The first stage of the surgery involves the surgeon creating a pocket within the existing scarred cornea and inserting the device (after removing a portion of the scarred cornea first). This is usually performed under general anesthetic. Approximately 3 months after this initial surgery, a second, smaller procedure is performed. At this stage, the scarred tissue that is covering the AlphaCor is removed to allow the light to enter.

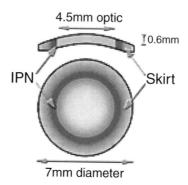

FIGURE 57: AlphaCor artificial cornea. Specifics of the outer skirt and the IPN are explained in the text.

Another type of artificial (nontissue) cornea is the Boston keratoprosthesis, manufactured by the Massachusetts Eye and Ear Hospital, which is affiliated to the Harvard Medical School. This device is a multipiece system as shown in Figure 58.

The Boston system appears to be similar to a bolt, washer and screw assembly, albeit biocompatible and sterilizable versions. The Boston keratoprosthesis has been under development since the 1960s and has gradually been perfected. It received FDA approval in 1992. About 600 implantations have been performed. It is one of the most commonly used keratoprosthesis in the United States. The keratoprosthesis is made of clear plastic with excellent tissue tolerance and optical properties. It consists of two parts but when fully assembled, it has the shape of a collar button (see Figure 58). The device is inserted into a corneal graft, which is then sutured into the patient's cloudy cornea. If the natural lens is in place, it is also removed. Finally, your

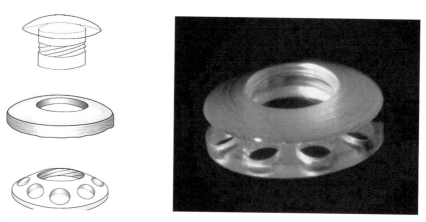

FIGURE 58: Boston keratoprosthesis consisting of three sections.

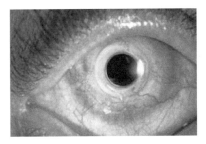

FIGURE 59: Boston keratoprosthesis 3 months after implantation.

physician may recommend that a soft contact lens may be applied to the surface. The Boston keratoprosthesis is shown implanted in a patient 3 months postsurgery in Figure 59.

Postoperative medication might be fluoroquinolone and vancomycin drops (14 mg/mL) once or twice daily. Prednisolone acetate 1% (a steroid) may also be used if needed. If pressure problems appear, steroids may be eliminated but the antibiotics should be used for life, on a daily basis. Soft contact lens should be worn around the clock on a long-term basis. The lens is very protective of the corneal tissue hydration.

Tissue-based artificial corneas consist of the following versions: 1) an artificial scaffold in which cells and nerves grow to form a natural corneal replacement, or 2) a tissue-engineered replacement that can replace a damaged cornea with a true tissue analog that will again grow into a finalized complete cornea. Scientists from the University of Ottawa created an artificial cornea that is grown around a "scaffold" of plastic and protein implanted into the eye. It regenerates the cells necessary to make a fully functioning cornea within a matter of weeks. So far the scaffold has been successfully tested only in pigs with corneal damage. Pigs given traditional corneal transplants showed no nerve regeneration in the weeks following transplantation. They reported, in the journal *Proceedings of the National Academy of Sciences*, that other corneal substitutes have been produced and tested, but that their implantable matrix performs as a physiologically functional tissue substitute and not simply as a prosthetic device (Griffith *et al.* 2002).

Another alternative is the true tissue-engineered replacement cornea. Human eye tissue has been grown in a laboratory and then successfully transplanted into patients. All the 14 patients in the University of California study had badly affected eyesight, but in 10 cases following treatment, vision was either restored or significantly improved. The scientists used the latest "bioengineering" techniques to grow corneal cells in the laboratory from a very small number supplied either by the patient, if there was one good eye, or by a related donor. Similar techniques are frequently used to "grow" new skin to lay over severe burns. The eye cells, called corneal stem cells, are naturally found just underneath the outside of cornea itself, which protects the delicate eye from damage. The stem cells mature into adult corneal cells that are needed to replace ageing corneal cells or repair injuries. Certain corneal injuries, such as burns—from fire, radiation, or chemicals—and some rare diseases and tumors can destroy the patient's stem cells. This means that the eye is no longer able to repair itself, and an accumulation of slight injuries destroys the eyesight. In these cases, a traditional corneal transplant, in which the outermost layer is taken from an organ donor, is simply not enough, as not enough stem cells are carried to replace those lost. The technique, perfected at the University of California Davis School of Medicine and Medical Center, harvests just a few stem cells under local anesthetic. These are converted into films just one cell thick in laboratory dishes, and placed on to a sterile membrane that gives them a framework on which to grow. A thicker layer is eventually produced, which is tough enough to be transplanted. The stem cells then mature into adult corneal cells so that sight can be restored. This work was initially reported by Schwab and Isserhoff (2000) followed by Han *et al.* (2002), and Schwab *et al.* (2000). Obviously, as was the case for the University of Ottowa study, the field of tissue and/or cellular based artificial corneas is still in its infancy. This is, however, the holy grail of corneal transplant, as is the case for organ transplant for other purposes—the ability to replace a failing organ with a replacement that looks, acts, and feels like the original.

REFERENCES

Aarts NL, Caffee CS. Manufacturer predicted and measured REAR values in adult hearing aid fitting: accuracy and clinical usefulness. *Int J Audiol* 2005 May;44(5):293–301.

Alio JL, Mulet ME, Gutierrez R, Galal A. Artisan toric phakic intraocular lens for correction of astigmatism. *J Refract Surg* 2005 Jul–Aug;21(4):324–31.

Beadle EA, McKinley DJ, Nikolopoulos TP, Brough J, O'Donoghue GM, Archbold SM. Long-term functional outcomes and academic-occupational status in implanted children after 10 to 14 years of cochlear implant use. *Otol Neurotol* 2005 Nov;26(6):1152–60.

Benedetti S, Casamenti V, Marcaccio L, Brogioni C, Assetto V. Correction of myopia of 7 to 24 diopters with the Artisan phakic intraocular lens: two-year follow-up. *J Refract Surg* 2005 Mar–Apr;21(2):116–26.

Bentler RA. Effectiveness of directional microphones and noise reduction schemes in hearing aids: a systematic review of the evidence. *J Am Acad Audiol* 2005 Jul–Aug;16(7): 473–84.

Berg AL, Herb A, Hurst M. Cochlear implants in children: ethics, informed consent, and parental decision making. *J Clin Ethics* 2005 Fall;16(3):239–50.

Biernath KR, Reefhuis J, Whitney CG, Mann EA, Costa P, Eichwald J, Boyle C. Bacterial meningitis among children with cochlear implants beyond 24 months after implantation. *Pediatrics* 2006 Jan 3.

Cakmak SS, Caca I, Unlu MK, Cakmak A, Olmez G, Sakalar YB. Surgical technique and postoperative complications in congenital cataract surgery. *Med Sci Monit* 2005 Dec 19;12(1):CR31–35.

Clark G, Shepherd R, Patrick J, Black R, Tong Y. Design and fabrication of the banded electrode array. *Ann NY Acad Sci* 1983;405:191–201.

Cohen-Mansfield J, Infeld DL. Hearing aids for nursing home residents: current policy and future needs. *Health Policy* 2005 Dec 30.

Collins JF, Gaster RN, Krol WF, VA Cooperative Cataract Study Group. Outcomes in patients having vitreous presentation during cataract surgery who lack capsular support for a nonsutured PC IOL. *Am J Ophthalmol* 2006 Jan;141(1):71–78.

Fabry D. Creating the evidence: lessons from cochlear implants. *J Am Acad Audiol* 2005 Jul–Aug;16(7):515–22.

Ferlito A, Arnold W, Rinaldo A, Niedermeyer HP, Bozorg Grayeli A, Devaney KO, McKenna MJ, Anniko M, Pulec JL, McCabe BF, van den Broek P, Shea JJ Jr. Viruses and otosclerosis: chance association or true causal link? *Acta Otolaryngol* 2003 Aug;123(6):741–6.

Flipsen P Jr, Colvard LG. Intelligibility of conversational speech produced by children with cochlear implants. *J Commun Disord* 2005 Dec 20.

Frazer RQ, Byron RT, Osborne PB, West KP. PMMA: an essential material in medicine and dentistry. *J Long Term Eff Med Implants* 2005;15(6):629–39.

Fu QJ, Chinchilla S, Nogaki G, Galvin JJ 3rd. Voice gender identification by cochlear implant users: the role of spectral and temporal resolution. *J Acoust Soc Am* 2005 Sep;118(3 Pt 1):1711–8.

Gordon KA, Tanaka S, Papsin BC. Atypical cortical responses underlie poor speech perception in children using cochlear implants. *Neuroreport* 2005 Dec 19;16(18):2041–5.

Griffith M, Hakim M, Shimmura S, Watsky MA, Li F, Carlsson D, Doillon CJ, Nakamura M, Suuronen E, Shinozaki N, Nakata K, Sheardown H. Artificial human corneas: scaffolds for transplantation and host regeneration. *Cornea* 2002 Oct;21(7 Suppl):S54–61.

Gustav Mueller H. Fitting hearing aids to adults using prescriptive methods: an evidence-based review of effectiveness. *J Am Acad Audiol* 2005 Jul–Aug;16(7):448–60.

Gustav Mueller H, Bentler RA. Fitting hearing aids using clinical measures of loudness discomfort levels: an evidence-based review of effectiveness. *J Am Acad Audiol* 2005 Jul–Aug;16(7):461–72.

Hachmeister JE. An abbreviated history of the ear: from Renaissance to present. *Yale J Biol Med* 2003 Mar 1;76(2):81–6.

Haensel J, Ilgner J, Chen YS, Thuermer C, Westhofen M. Speech perception in elderly patients following cochlear implantation. *Acta Otolaryngol* 2005 Dec;125(12):1272–6.

Han B, Schwab IR, Madsen TK, Isseroff RR. A fibrin-based bioengineered ocular surface with human corneal epithelial stem cells. *Cornea* 2002 Jul;21(5):505–10.

Hay-McCutcheon MJ, Pisoni DB, Kirk KI. Audiovisual speech perception in elderly cochlear implant recipients. *Laryngoscope* 2005 Oct;115(10):1887–94.

Heermann H. Johannes Kessel and the history of endaural surgery. *Arch Otolaryngol* 1969 Nov;90(5):652–8.

Henkin Y, Kaplan-Neeman R, Kronenberg J, Migirov L, Hildesheimer M, Muchnik C. Electrical stimulation levels and electrode impedance values in children using the Med-El Combi 40+ cochlear implant: a one year follow-up. *J Basic Clin Physiol Pharmacol* 2005;16(2–3):127–37.

Higgins MB, McCleary EA, Ide-Helvie DL, Carney AE. Speech and voice physiology of children who are hard of hearing. *Ear Hear* 2005 Dec;26(6):546–58.

Hochmair-Desoyer I, Hochmier E, Burian K. Design and fabrication of multiwire scala tympani electrodes. *Ann NY Acad Sci* 1983; 405:173–82.

Hong RS, Rubinstein JT. Conditioning pulse trains in cochlear implants: effects on loudness growth. *Otol Neurotol* 2006 Jan;27(1):50–56.

House HP, Wullstein H, Shea JJ Jr, Derlacki EL, Schuknecht H, Foller EP Jr, Juers Al. Techniques of stapes mobilization. *Arch Otolaryngol* 1960 Feb;71:338–53.

House W, Urban J. Long term results of electrode implantation and electronic stimulation of the cochlea in man. *Ann Otol Rhinol Laryngol* 1973;82:504–17.

Hyde M, Power D. Some ethical dimensions of cochlear implantation for deaf children and their families. *J Deaf Stud Deaf Educ* 2006 Winter;11(1):102–11.

Karosi T, Konya J, Petko M, Sziklai I. Histologic otosclerosis is associated with the presence of measles virus in the stapes footplate. *Otol Neurotol* 2005 Nov;26(6):1128–33.

Karosi T, Konya J, Szabo LZ, Sziklai I. Measles virus prevalence in otosclerotic stapes footplate samples. *Otol Neurotol* 2004 Jul;25(4):451–6.

Killion MC, Gudmundsen GI. Fitting hearing aids using clinical prefitting speech measures: an evidence-based review. *J Am Acad Audiol* 2005 Jul–Aug;16(7):439–47.

Laneau J, Wouters J, Moonen M. Improved music perception with explicit pitch coding in cochlear implants. *Audiol Neurootol* 2006 Jan–Feb;11(1):38–52.doi:10.1159/000088853

Leccisotti A, Fields SV. Clinical results of ZSAL-4 angle-supported phakic intraocular lenses in 190 myopic eyes. *J Cataract Refract Surg* 2005 Feb;31(2):318–23. doi:10.1016/j.jcrs.2004.04.051

Lempert J, Meltzer PE, Rambo JH, Wever EG. The effects of injury to the lateral semicircular canal. *Trans Am Acad Ophthalmol Otolaryngol* 1956 Sep–Oct;60(5):718–27.

Lewis MS, Valente M, Horn JE, Crandell C. The effect of hearing aids and frequency modulation technology on results from the communication profile for the hearing impaired. *J Am Acad Audiol* 2005 Apr;16(4):250–61.

Lifshitz T, Levy J, Aizenman I, Klemperer I, Levinger S. Artisan phakic intraocular lens for correcting high myopia. *Int Ophthalmol* 2004 Jul;25(4):233–8. Epub 2005 Sep 29. doi:10.1007/s10792-005-5016-2

Loizou PC. Introduction to cochlear implants. *IEEE Signal Processing Mag* 1998 Sep:101–130.

Lovisolo CF, Reinstein DZ. Phakic intraocular lenses. *Surv Ophthalmol* 2005 Nov–Dec;50(6):549–87.doi:10.1016/j.survophthal.2005.08.011

Martin RG, Sanders DR. A comparison of higher order aberrations following implantation of four foldable intraocular lens designs. *J Refract Surg* 2005 Nov–Dec;21(6):716–21.

Merzenich M, White M. Cochlear implant – the interface problem, in *Functional Electrical Simulation: Applications in Neural Protheses* (F. Hambrecht and J. Reswick eds.). New York: Marcel Dekker, 1977, pp. 321–340.

Moore BC, Marriage J, Alcantara J, Glasberg BR. Comparison of two adaptive procedures for fitting a multi-channel compression hearing aid. *Int J Audiol* 2005 Jun;44(6):345–57. doi:10.1080/14992020500060198

Munson B, Nelson PB. Phonetic identification in quiet and in noise by listeners with cochlear implants. *J Acoust Soc Am* 2005 Oct;118(4):2607–17.doi:10.1121/1.2005887

Myers EN, Ishiyama E, Heisse JW Jr. Histology of a successful stapes mobilization. *Ann Otol Rhinol Laryngol* 1970 Apr;79(2):321–30.

Myers EN, Myers D. Stapedectomy in advanced otosclerosis: a temporal bone report. *J Laryngol Otol* 1968 Jun;82(6):557–64.

Niedermeyer HP, Arnold W, Schuster M, Baumann C, Kramer J, Neubert WJ, Sedlmeier R. Persistent measles virus infection and otosclerosis. *Ann Otol Rhinol Laryngol* 2001 Oct;110(10):897–903.

Oshika T, Kawana K, Hiraoka T, Kaji Y, Kiuchi T. Ocular higher-order wavefront aberration caused by major tilting of intraocular lens. *Am J Ophthalmol* 2005 Oct;140(4):744–6. doi:10.1016/j.ajo.2005.04.026

Parsons C, Jones DS, Gorman SP. The intraocular lens: challenges in the prevention and therapy of infectious endophthalmitis and posterior capsular opacification. *Expert Rev Med Devices* 2005 Mar;2(2):161–73.doi:10.1586/17434440.2.2.161

Perkins RC. Laser stepedotomy for otosclerosis. *Laryngoscope* 1980 Feb;90(2):228–40.

Reber MB, Kompis M. Acclimatization in first-time hearing aid users using three different fitting protocols. *Auris Nasus Larynx* 2005 Dec;32(4):345–51.doi:10.1016/j.anl.2005.05.008

Reese JL, Hnath-Chisolm T. Recognition of hearing aid orientation content by first-time users. *Am J Audiol* 2005 Jun;14(1):94–104.doi:10.1044/1059-0889(2005/009)

Ricketts TA, Hornsby BW. Sound quality measures for speech in noise through a commercial hearing aid implementing digital noise reduction. *J Am Acad Audiol* 2005 May;16(5):270–7.

Schuknecht HF, Frederick L. Jack (1861–1951). *Arch Otolaryngol* 1968 Mar;87(3):328–32.

Schwab IR, Isseroff RR. Bioengineered corneas – the promise and the challenge. *N Engl J Med.* 2000 Jul 13;343(2):136–8.doi:10.1056/NEJM200007133430211

Schwab IR, Reyes M, Isseroff RR. Successful transplantation of bioengineered tissue replacements in patients with ocular surface disease. *Am J Ophthalmol* 2000 Oct;130(4):543–4. doi:10.1016/S0002-9394(00)00747-9

Shampo MA, Kyle RA. Georg von Bekesy – audiology and the cochlea. *Mayo Clin Proc* 1993 Jul;68(7):706.

Shea JJ Jr. A personal history of stapedectomy. *Am J Otol* 1998 Sep;19(5 Suppl):S2–12.

Shea MC. Stapes fixation in chronic middle ear disease. *Laryngoscope* 1976 Feb;86(2):230–2.

Smith AF, Lafuma A, Berdeaux G, Berto P, Brueggenjuergen B, Magaz S, Auffarth GK, Brezin A, Caporossi A, Mendicute J. Cost-effectiveness analysis of PMMA, silicone, or acrylic intra-ocular lenses in cataract surgery in four European countries. *Ophthalmic Epidemiol* 2005 Oct;12(5):343–51.doi:10.1080/09286580500180598

Sridhar MS, Majji AB, Vaddavalli PK. Severe inflammation following iris fixated anterior chamber phakic intraocular lens for myopia. *Eye* 2005 Oct 21.

Stickney GS, Loizou PC, Mishra LN, Assmann PF, Shannon RV, Opie JM. Effects of electrode design and configuration on channel interactions. *Hear Res* 2005 Dec 7.

Stickney GS, Nie K, Zeng FG. Contribution of frequency modulation to speech recognition in noise. *J Acoust Soc Am* 2005 Oct;118(4):2412–20.doi:10.1121/1.2031967

Tassignon MJ, De Groot V, Van Tenten Y. Searching the way out for posterior capsule opacification. *Verh K Acad Geneeskd Belg* 2005;67(4):277–88.

Tognetto D, Sanguinetti G, Ravalico G. Tissue reaction to hydrophilic intraocular lenses. *Expert Rev Med Devices* 2005 Jan;2(1):57–60.doi:10.1586/17434440.2.1.57

Tomblin JB, Barker BA, Spencer LJ, Zhang X, Gantz BJ. The effect of age at cochlear implant initial stimulation on expressive language growth in infants and toddlers. *J Speech Lang Hear Res* 2005 Aug;48(4):853–67.doi:10.1044/1092-4388(2005/059)

Tyler R. Open-set recognition with the 3M/Vienna single-channel cochlear implant. *Arch Otolaryngol Head Neck Surg* 1998;114:1123–6.

Uriarte M, Denzin L, Dunstan A, Sellars J, Hickson L. Measuring hearing aid outcomes using the Satisfaction with Amplification in Daily Life (SADL) questionnaire: Australian data. *J Am Acad Audiol* 2005 Jun;16(6):383–402.

van den Honert C, Stypulkowski P. Single fiber mapping of spatial excitation patterns in the electrically stimulated nerve. *Hear Res* 1987;29:195–206.doi:10.1016/0378-5955(87)90167-5

van Hooren SA, Anteunis LJ, Valentijn SA, Bosma H, Ponds RW, Jolles J, van Boxtel MP. Does cognitive function in older adults with hearing impairment improve by hearing aid use? *Int J Audiol* 2005 May;44(5):265–71.doi:10.1080/14992020500060370

Vermeire K, Brokx JP, Wuyts FL, Cochet E, Hofkens A, De Bodt M, Van de Heyning PH. Good speech recognition and quality-of-life scores after cochlear implantation in patients with DFNA9. *Otol Neurotol* 2006 Jan;27(1):44–49. doi:10.1097/01.mao.0000187240.33712.01

Vuorialho A, Sorri M, Nuojua I, Muhli A. Changes in hearing aid use over the past 20 years. *Eur Arch Otorhinolaryngol* 2005 Nov 9.

Waltzman SB, Cohen NL, Shapiro WH. The benefits of cochlear implantation in the geriatric population. *Otolaryngol Head Neck Surg* 1993 Apr;108(4):329–33.

Werner L, Hunter B, Stevens S, Chew JJ, Mamalis N. Role of silicon contamination on calcification of hydrophilic acrylic intraocular lenses. *Am J Ophthalmol* 2006 Jan;141(1):35–43. doi:10.1016/j.ajo.2005.08.045

Wilson B, Lawson D, Zerbi M. Advances in coding strategies for cochlear implants. *Adv Otolaryngol Head Neck Surg* 1995;9:105–29.

Xu L, Zwolan TA, Thompson CS, Pfingst BE. Efficacy of a cochlear implant simultaneous analog stimulation strategy coupled with a monopolar electrode configuration. *Ann Otol Rhinol Laryngol* 2005 Nov;114(11):886–93.

Author Biography

Gerald E. Miller is the Department Head of Biomedical Engineering at Virginia Commonwealth University as well as a Professor of Physiology, Professor of Cardiology, and a professor of Physical Medicine & Rehabilitation. The Biomedical Engineering Department is administered through the School of Engineering, but is situated within the Medical College of Virginia campus of the university, which contains the nations fourth largest academic medical complex. The Biomedical Engineering Department maintains graduate degrees leading to the M.S. and Ph.D. with 9 primary and 60 affiliate faculty and 65 graduate students as well as Virginia's first undergraduate BME program with an enrollment of 180. Dr. Miller received the B.S. in Aerospace Engineering from the Pennsylvania State University in 1971, the M.S. in Bioengineering from the Pennsylvania State University in 1975, and the Ph.D. in Bioengineering from the Pennsylvania State University in 1978. He is a member of Phi Kappa Phi as well as a Fellow of both ASME and AIMBE. His research activities include Rehabilitation Engineering, Physiological Fluid Mechanics, Artificial Internal Organs, Epilepsy Genesis, and the use of physiological signals in the control of mechanical systems. He has developed a multiple disk, centrifugal artificial ventricle, an early warning detection system for epileptic seizures, and has developed a protocol for prevention and treatment of decubitus ulcers. He has also developed a real time, hands free, PC based voice controller with infrared activated grippers for robots which can be easily implemented by disabled individuals. He has been an invited speaker at several international conferences in the area of fluid mechanics, artificial organs, and man-machine interfaces.

Printed in the United States
by Baker & Taylor Publisher Services